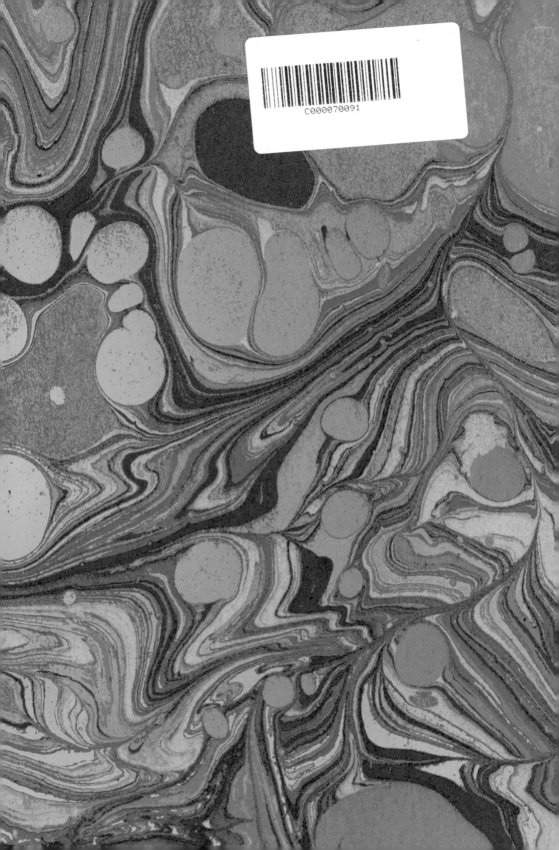

PSYCHONAUTS

PSYCHONAUTS

Drugs and the Making of the Modern Mind

MIKE JAY

YALE UNIVERSITY PRESS
NEW HAVEN AND LONDON

For information about this and other Yale University Press publications, please contact:
U.S. Office: sales.press@yale.edu   yalebooks.com
Europe Office: sales@yaleup.co.uk   yalebooks.co.uk

Set in Adobe Garamond Pro by IDSUK (DataConnection) Ltd
Printed in Great Britain by TJ Books, Padstow, Cornwall

Library of Congress Control Number: 2022948975

ISBN 978-0-300-25794-6

A catalogue record for this book is available from the British Library.

10 9 8 7 6 5 4 3 2 1

CONTENTS

ILLUSTRATIONS

BEFORE DRUGS

In 1992, a paper published in the scholarly journal *Ancient Mesoamerica* lit the fuse of a media drug panic. Two eminent scientists and bestselling authors, the anthropologist Wade Davis and the physician Andrew Weil, reported that they had resolved a long-standing puzzle in Mexican prehistory. It had long been noted that the tropical cane toad (*Bufo marinus*) appeared to have been important to the ancient Olmec and Mayan cultures: it was commonly represented on their ceramics and its bones were found in great abundance at several ritual sites. It was also well known that the toad's parotid glands behind its head secreted a mix of toxins that included a powerful hallucinogenic compound, bufotenine. Was it possible, as some anthropologists had suggested, that toad secretions had been used in ancient Mesoamerica as a ritual intoxicant?

Davis and Weil were both well versed in the history, ethnography and biology of mind-altering plants and fungi in Central America, and were fascinated by the possibility that a drug extracted from an amphibian might have been used alongside them. But 'the central weakness of the hallucinogen hypothesis', as Davis wrote, 'seemed to be the inability of proponents to demonstrate how any preparation of *Bufo marinus* could be safely consumed', since the bufotenine in the cane toad's gland occurs in combination with potent and potentially fatal cardiac toxins.[1] Their researches had suggested a possible solution: a related species of toad with a biochemistry that made it a more plausible candidate. *Bufo alvarius*, a toad native to the Sonora desert that straddles the US–Mexico border, was unique among its genus in possessing an enzyme in its poison – O-methyl transferase – that converts bufotenine into one of the most powerful hallucinogens found in nature, 5-MEO-DMT. If the venom of *B. alvarius* was smoked, as opposed to swallowed, the toxins would be broken down and the secretions should, in theory, be both safe and powerfully psychoactive.

The pair concluded that there was only one way to test their hypothesis. Experiments on laboratory animals were worthless, since there was no way to observe or verify whether white mice were hallucinating. They had, however, come across a 16-page pamphlet published by an underground press in 1984 that described in detail how to squeeze the parotid glands of a toad to extract its venom and allow it to dry into a smokable resin, and they followed its directions scrupulously.[2] They placed a chip of resin the size of a matchhead in a pipe and inhaled deeply, and within 15 seconds they had their answer. In their co-authored paper, Davis described the experience in language that blended clinical self-observation with vivid metaphor:

> Shortly after my inhalation I experienced warm flushing sensations, a sense of wonder and well-being, strong auditory sensations, which included an insect-cicada sound that ran across my mind and seemed to link my body to the earth . . . Strong visual hallucinations in orblike brilliance, diamond patterns that undulated across my visual field. The experience was in every sense pleasant, with no disturbing physical sensations, no nausea, perhaps a slight sense of increased heart rate.[3]

Davis and Weil's paper concluded that, since there were well-established trade routes between the Sonora desert and the Maya lowlands to the south, it was possible that this dried venom had been traded between them. It was an ingenious solution to a long-standing archaeological conundrum, but the response to their paper was one of shock and outrage. Members of their scientific community were scandalised that the authors had experimented on themselves with an untested and potentially dangerous drug, and had reported the experience in glowingly positive terms. Davis and Weil submitted their findings to the leading journals *Science* and *Nature*, but the article was rejected by both. When it eventually appeared in *Ancient Mesoamerica*, it was denounced as unethical and dangerous. One reviewer dismissed their experiment on the grounds that there had been no control group; another wrote that it amounted to 'little more than an endorsement for abuse of an hallucinogenic material with potentially deadly side effects'.[4]

The authors' drug experiment became the story. It was picked up by the leading newspaper in Davis's home nation of Canada, the *Toronto Globe and Mail*, under the headline 'Taking Their Own Medicine'. The article quoted the chairman of the University of Toronto's medical ethics review committee: 'We discourage self-experimentation very strongly.'[5] The press was scandalised by Andrew Weil's statement, 'I think the kind of inspiration which comes from experience makes for the best kind of science.'[6] The *Wall Street Journal* ran a front-page story connecting Davis and Weil's work to a shadowy subculture of toad lickers and smokers inspired by the same underground pamphlet. Rumours of toad-licking or smoking proved to be irresistible copy for a mass media that reflexively treated all drug use as delinquent or dangerous. In 1994, this *ne plus ultra* of depraved addictions found its public face. A naturalist and elementary school teacher in California named Bob Shepard, together with his wife Connie, were arrested by drug enforcement agents and their four pet specimens of *Bufo alvarius* confiscated after they confessed to smoking their secretions. Bob was fired from his job and ordered into drug rehabilitation.

* * *

It was around this time that I started writing about drugs, and I remember the media landscape well. The subject was framed as a medical, social and criminal problem. Institutional scientific research

3

into mind-altering drugs, or 'drugs of abuse', was focused exclusively on their negative effects, which supplied the media with a constant drip of scare stories: crack babies, ecstasy deaths, cannabis-induced psychosis. The mission statement of the National Institute of Drug Abuse, the US federal body that leads the global research field, was 'bringing the power of science to bear on drug abuse and addiction'.[7] Studies of their potential benefits were vanishingly rare. Scientists who spoke publicly about their drug use were – like doctors, lawyers, politicians and other professionals – taking a grave risk with their reputations and careers. Scientific self-experiments, as Davis and Weil discovered, were taboo. To admit to experience of drugs was to be marginalised within the public debate as a 'user': not an expert witness, but part of the problem. When newspapers or TV programmes required a 'drugs expert', they contacted an addiction psychiatrist or state-funded anti-drugs campaigner.

I was an early adopter of the internet, and was fascinated by the profusion of drug-themed bulletin boards and alt.drugs newsgroups that were among the first information exchanges to colonise the online ecosystem. Recipes for synthesising LSD circulated along with street prices for cannabis from Canada to Cambodia, suppressed pro-legalisation dossiers from the World Health Organization and discussions about how ecstasy interacted with anti-depressant medications. (The only topic off limits was actually buying and selling drugs, which was firmly stamped on by site moderators.) All this was unthinkable in the mainstream media, where every drug story was obliged to carry a health warning and a moral message. But the early internet was a far more accurate reflection of the globalised drug culture that proliferated in the streets, clubs and festivals of the 1990s, manifesting in vast, often illegal 'acid house parties' or raves fuelled by MDMA (ecstasy) and a rapidly expanding illicit pharmacopoeia. In the newspapers, this subculture was typically reported under headlines such as 'Dance of Death'; but online – at this point, a strange new world of green-on-black text, blinking cursors and MS-DOS prompts – drugs were simply

4

another specialist topic, discussed with fellow enthusiasts as one might share an interest in science, travel or literature.

I sold a few pieces about this scene to the style and youth mags of the day, but even here it was regarded as a marginal topic: drugs were a disreputable subject that drew unwelcome attention, and the internet was strictly for geeks.[8] 'Drug culture', for the popular media, was an oxymoron. It had its own domain, such as it was, in the margins: an underground press, mostly based in the USA, specialising in cannabis-growing guides and reprinted texts by hippie-era luminaries like Timothy Leary, and a samizdat, often hand-distributed network of drug-user 'zines, freesheets and flyers.[9] Beyond these outlaw enclaves, the obvious-seeming questions – why so many people took drugs, what their effects were, and how they influenced the cultures in which they circulated – were rarely posed. When they were, the drug experts typically responded with one of two answers: 'peer pressure' or 'addiction'. In both cases, the remedy was more anti-drug messages and harsher criminal penalties.

This made me curious about the history of drugs. Where had all these mind-altering substances arrived from in the first place, and who had been the first people to try them? What had their experiences been, and how had drugs come to establish themselves in the modern world?

The academic history of drugs at that time was a small specialist field, largely focused on criminology and medical pathology: institutional studies of the development of law enforcement and illegal trafficking, and medical histories of addiction theory and public health interventions. What was lacking from almost all of it was any attention to, or curiosity about, the effects of drugs on the mind – surely, I felt, the most significant aspect of mind-altering drugs. It was usually clear from the opening page that the authors had no experience of drugs themselves, and no more interest in acquiring it than epidemic historians had in contracting smallpox or bubonic plague. Those most engaged with the drug experience were literary studies of the opiated Romantic poets, such as Thomas De Quincey, although these never adduced personal experience and rarely questioned or deviated from the assumption that drugs were – as Charles Baudelaire had famously claimed – 'a forbidden game', in which any fleeting insights or epiphanies came at a punishing price.[10]

I was surprised to discover that the War on Drugs, as it is popularly known, was a relatively recent development, its origins at that time within living memory. To look back beyond the early twentieth century, when drugs were criminalised, was to enter a quite different world: one in which even the term 'drugs' no longer applied. The word had existed, of course – it entered the English language in medieval times, probably from the Dutch root denoting 'dried goods', as supplied by apothecaries – but only in its broader sense, a general term for all medications used by doctors. This meaning still exists, for example in 'drugstore'; but 'drugs' in its more specific sense was a creation of the twentieth century. Before then, mind-altering, intoxicating or psychoactive drugs were a capacious and loosely linked family that spanned household remedies, pharmacy products, scientific research chemicals and herbal intoxicants. There were some general statutory controls on poisons, a category that often included opium alongside arsenic, strychnine and cyanide, but many of today's illicit substances, from hashish to cocaine to heroin, were freely available over the pharmacy counter, sold as part of the general array of analgesics, stimulants and sedatives.

In this era, doctors and scientists routinely experimented with drugs on themselves. By the time of Wade Davis and Andrew Weil's experiment with toad secretions, it had apparently been forgotten, even by their scientific colleagues, that the practice had ever been acceptable, let alone a standard procedure. Yet it had been obvious ever since the scientific revolution of the seventeenth century that this was the most satisfactory method for investigating drugs that change perceptions, moods and consciousness. By the nineteenth century it was a practice with well-recognised protocols and reporting conventions. In an age of experiment, when scientific and medical breakthroughs often involved personal risk, sampling an unknown drug was seen as no more reckless than the anatomical experiments, electric shocks or toxic exposure to which the researchers of the day routinely subjected themselves.

Naturally, most scientists took care to minimise these risks: in the case of drugs, by practices such as careful preparatory research, working up incrementally from small doses, using sterile needles and pure solutions. At the same time, many were powerfully drawn to the thrill of exceeding the limits of the known, often with dramatic consequences. Reckless drug experiments and descents into madness were at the core of many of the era's most popular fictions. Self-experimenters were celebrated as popular icons: the British chemist Humphry Davy, for example, launched himself as the scientific hero of his generation with thrilling poetic descriptions of inhaling huge quantities of nitrous oxide and exploring unknown realms of disembodied consciousness. The site of his experiments in Bristol is today marked with a bronze plaque, and his giant statue looms over the main street of his home town, Penzance.

Experiments of this kind were not confined to science. The nineteenth century was an age in which ideas and discoveries flowed easily between medicine and science, literature and art, philosophy and spiritual exploration. Drugs – I'll persist with the anachronistic but convenient term – tended to generate experiences that transcended and dissolved these boundaries. Doctors and scientists were trained in careful self-observation, and drugs that produced novel sensations and states of consciousness were a spur to exercise their powers of description to the fullest. Not infrequently they were also poets or novelists, and brought these skills to their medical observation. Before the nineteenth century, medical case histories had typically been recorded as concise notes of diagnosis and family background; now they were increasingly recast as discursive narratives of the inner life, adopting stream-of-consciousness styles or multiple viewpoints. Many writers and other artists, in turn, avidly followed the scientific discoveries of the day and were among the first experimenters with novel drugs. Authors of what Émile Zola termed the 'experimental novel' turned

7

a scientific gaze on the workings of society, and artists drew on drug experiences, along with the latest findings in perceptual psychology, to explode the conventional rules of painting, colour, shape and movement.

At the time, there was no specific term for those who used drugs to explore the mind. It was only in 1949 that one was coined, by the German writer Ernst Jünger in his novel *Heliopolis*, about an oppressive futuristic city where a rebellious scientist finds freedom through his drug-induced inner journeys. Jünger invented the term 'psychonaut' to describe a character who 'captured dreams, just as others seem to pursue butterflies with nets' and 'went on voyages of discovery in the universe of his brain'.[11] Jünger was a spiritual mentor to Albert Hofmann, the discoverer of LSD, who helped to popularise the term, and it came into currency through the psychedelic counterculture of the 1960s; but it serves as a useful collective term for a previous generation whose inner journeys had been largely forgotten by the time it was coined. These days it suggests a renegade working outside the boundaries of institutional science; the early psychonauts also included plenty of renegades, autodidacts, bohemians and mystics, but they experimented alongside leading scientists, university professors, doctors, surgeons, business leaders, philosophers and pillars of the literary establishment.

The writings of these self-experimenters were a source of intense fascination to the wider public. Had Wade Davis and Andrew Weil written up their experiment a century earlier, it might easily have found a home in one of the mass-market journals and magazines that routinely published such marvels of mental science alongside explorers' descriptions of far-flung landscapes, sensational tales of true crime, or the latest instalment of a gothic mystery novel. Their accounts posed far-reaching questions – chemical, medical and philosophical – about the human condition and the limits of the mind. Could a substance such as cocaine actually produce superhuman energy or enhanced cognitive powers? Did the newly discovered anaesthetic gases and vapours allow the mind access to an unexplored dimension beyond the limits of the body, and what was to be found there? Could the magical herbs and potions of ancient myths or non-Western cultures be used to access prophetic visions, telepathy, hypnotic powers or contact with the spirit world?

All these questions fed a growing obsession with the hidden regions of the mind that reached a peak of intensity in the final decades of the nineteenth century. The new discipline of psychology was making giant strides in teasing apart the processes of thought, attention, perception and sensation, and mapping the complex interplay between eye and brain. Many of its practitioners were fascinated by drugs that allowed them to observe the mind as it made sense of strange new stimuli, and even manufactured hallucinatory realities. For psychiatrists and neurologists, puzzling over uncanny cases of amnesia, somnambulism, hysteria or dual personality, drugs became a valuable tool for accessing the hidden mind: unlike the tantalising phenomena witnessed under hypnosis or in spirit séances, they could be administered in repeatable experiments and measurable doses. For writers and artists, they were the portal to an inner world where dream and reality coexisted. For spiritualists and psychical researchers, the experiences they engendered were evidence for the reality of other worlds or dimensions; practitioners of the occult arts incorporated them into lucid dreaming, astral travel and ceremonial magic. As religion retreated before the rising tide of science, ineffable experiences that would previously have been accepted as divinely inspired took on new meanings, both profane and sacred.

Drugs arrived in the twentieth century as miracles of science that held the promise of transforming the human condition and reshaping its future. They allowed their subjects to escape the straitjacket of time and space, and to examine reality from novel and multiple viewpoints. They expanded the domain of subjectivity, allowing their subjects to inhabit private worlds of imagination, thought and sensation, and extending the reach of spiritual exploration and artistic creativity. In doing so, they fed into a wider intellectual trend away from a single, settled and hierarchical idea of culture – famously defined by Matthew Arnold in 1869 as 'the best which has been thought and said'[12] – towards a plurality of perspectives and mental states, a multiplicity of subcultures and the fractured, kaleidoscopic 'stream of consciousness' that would preoccupy the next generation of writers and artists.

The dangers represented by drugs were no less compelling. Long before they were made illegal, they encapsulated the terrors of modernity as vividly as its possibilities. From the 1870s, when doctors began to diagnose the condition of 'morbid craving' or addiction to opiates and cocaine, it was recognised as a baneful consequence of the modern freedom to indulge in consumption without limits in an age of ever more pure and potent drugs. For many, the most troubling drug of all was alcohol, particularly in the form of cheap distilled spirits, but by 1900 the temperance movement also included consumer groups lobbying for the removal of opiates and cocaine from patent medicines, which often contained them unadvertised and in liberal doses, their bitterness masked by sugary syrups.

Beyond their medical dangers, drugs provoked existential questions about how a modern civilised society could function if their use continued to spread. In the eyes of their detractors, those under their influence were egotists, indulging narcissistic and delusional fantasies at the expense of a consensual social fabric. The freedom to choose one's mode of consciousness, from this standpoint, was an open invitation to solipsism, irresponsibility and vice. The pleasure that drugs conferred on their subjects was morally corrosive: an unearned and empty chemical high, destructive to the work ethic and the social compact of shared responsibility. When the term 'drug' emerged in its specialised sense at the beginning of the twentieth century, it was freighted with negative connotations. 'Drugs', in the new usage, denoted dangerous substances only to be administered by a medical professional. Used irresponsibly, they preyed on the inadequacies of their subjects, seizing on pathological tendencies or hereditary weakness to erode their will-power, health and moral agency. Medical and political reformers often included them among the 'degenerate habits' of 'inferior races' that impeded the progress of modern, Western civilisation. By the 1920s, after they were brought under legal controls, the term 'drugs' had taken on the cluster of pejorative meanings – dangerous, foreign, criminal – with which the term is still freighted today.

During the twentieth century these associations hardened, reinforced by the steady march of international anti-drug laws; and in the process, the exhilarating sense of possibility that mind-expanding drugs had represented in an earlier era receded from view. The 'drugs' most frequently recalled from the previous era were opium and its more potent derivatives, morphine and heroin. These had indeed been the most widely used in nineteenth-century medicine and the most commonly referenced in its literature; they were also the most problematic, and the most prominent in the arguments that led to the twentieth century's drug control regime. By the late nineteenth century, however, many other drugs offered expansive possibilities for exploring the mind: the unprecedentedly powerful euphoria and mental stimulation of cocaine; the cosmic epiphanies of nitrous oxide, ether or chloroform; the visionary intoxication of cannabis and other plants long familiar in distant corners of the globe.

Much of the self-experimental literature generated by these substances was forgotten during the twentieth century, but a few episodes remained too conspicuous to ignore. Why, for example, had Sigmund Freud been such an enthusiastic champion of cocaine? And why did William James continue to insist that the insights he gained from inhaling nitrous oxide – laughing gas – were central to his understanding of consciousness and mystical experience? In most twentieth-century treatments of these towering figures in the discovery of the modern mind, their drug experiments are dismissed with embarrassment (or latched upon with glee by their detractors) as youthful escapades or foolish errors of judgement. An admiring survey of William James's career published in 1948 dismissed his nitrous oxide epiphany as 'the child playing with matches, or irreverently mocking the devout', and relegated it to a chapter titled 'Morbid Traits'.[13] Freud himself suppressed his work on cocaine after the drug was criminalised, omitting his papers on it from his bibliography; his biographer Ernest Jones followed him in minimising the episode in his authorised three-volume life and works, and in private judged it an aberration: 'he was only interested in the magical effect of the drug, of which he took too much himself'.[14]

Retrospective explanations of this kind tell us more about the prejudices of the century in which they were written than about the events they describe. Freud and James were working in an era where such experiments were not unusual or eccentric; many of their contemporaries did the same, and indeed considered self-experiment as a mark of seriousness or professional dedication. I want to restore this missing context by setting the drug experiences and discoveries of such major figures alongside those of their less famous contemporaries, both in the mind sciences and in the wider intellectual worlds – from literature to art history, philosophy to spiritualism – across which thinkers such as Freud and James ranged.

This nineteenth-century body of drug writing feels much more accessible in the twenty-first century than it did in the twentieth. The lines are easily drawn to today's fascination with psychedelics and other mental enhancers, and their promises to free our minds from the over-rigid strictures of modern life. The early psychonauts remain of their time in some obvious respects: they were almost all educated white males, a reflection of that group's domination of nineteenth-century science, for which reason I have made the most of female, working-class, non-white and non-Western voices where I have found them. In other respects, however, their diversity is striking, and their endeavours amount to a remarkable, and remarkably under-studied, episode in Western intellectual history.

* * *

In the twenty-first century, the time is ripe for its rediscovery. Over the thirty years since I began writing about drugs, our attitude to them has changed beyond recognition. Take the example of smoking Sonoran desert toad secretions, which has made the transit from the far reaches of drug-addled degeneracy to become the latest, much-hyped arrival in the crowded marketplace of psychedelic therapies. Clinical studies have hailed the promise of 'the toad' for the treatment of depression, anxiety and stress, and earned its active ingredient, 5-MEO-DMT, the unlikely epithet of 'the God molecule'.[15] Global influencers and opinion formers, from Burning Man to Davos, have become its evangelists. In 2019 Mike Tyson told Joe Rogan and his 200 million podcast listeners that smoking the toad had led him to insights that turned his life around. Rogan himself has hailed it as the equivalent of fifteen years of psychotherapy. In 2021 Hunter Biden, the son of the US president, published a memoir of his recovery from substance abuse in which he credited smoking toad, under the guidance of a spiritual healer at a beach house in Mexico, with keeping him clean for a year: 'It was a profound experience,' Biden wrote; 'it connected me in a vividly renewed way to everyone in my life.'[16] So many self-styled shamans now offer it as a commercial therapy or sacred medicine across the deserts of south-west USA and northern Mexico that conservationists and indigenous groups are concerned the Sonoran desert toad is being exploited unsustainably across its fragile natural habitat. In response, a process has been developed to produce the compound in synthetic form for clinical trials.[17]

The experience produced by smoking toad is famously ineffable, but Michael Pollan took up the challenge of describing it in his best-selling survey of the new psychedelic landscape, *How to Change your Mind* (2018). The task is the same that nineteenth-century experimenters with nitrous oxide or ether faced when struggling to put into words a disembodied journey that might equally have lasted for a second or for eternity. Pollan relates the overwhelming experience in carefully considered prose that recalls the classic accounts of Humphry Davy or William James:

> I have no memory of ever having exhaled, or of being lowered onto the mattress and covered with a blanket. All at once I felt a tremendous rush of energy fill my head accompanied by a punishing roar . . . 'I' was no more, blasted to a confetti cloud by an explosive force I could no longer locate in my head, because it had exploded that too, expanding to become all that there was. Whatever this was, it was not a hallucination. A hallucination implies a reality and a point of reference and an entity to have it. None of those things remained.[18]

A generation ago, the same self-experiment by Wade Davis and Andrew Weil was condemned as an irresponsible breach of scientific protocol. Today, as billions of research dollars pour into medical psychedelics and cognitive enhancement, it is time to recognise that our new-found fascination with drugs and the exploration of consciousness has a deeper history than we recognise. Many of its advocates and practitioners conceive it as entirely novel, a product of twenty-first-century neuroscience and perhaps the prelude to a neurochemically transformed post-human future. For others, it finds a charter in non-Western cultures: traditional Asian practices of meditation and mindfulness, or the New World's shamanism and indigenous plant wisdom. (I have written elsewhere about some of these global traditions, but my focus in this book is on the modern West.) These narratives are all underpinned by the assumption that the Western engagement with drug experiences is shallow, dating no further back than the 1960s and that decade's countercultural embrace of transcendence and self-actualisation. A longer view, however, reveals that we are reigniting an enduring fascination with drugs and the mind that was a hallmark of Western modernity a century ago, but disappeared from our collective memory during the drug-averse twentieth century.

Drugs today are both fetishised and demonised, and our conflicted attitudes towards them reflect a fault-line between competing versions of modernity. 'Modern' is not a term that can be claimed by one particular era, or by a single and uncontested set of values. Ever since the concept emerged during the Renaissance, when it described the new stage of civilisation that was succeeding the classical world and the Dark Ages, it has been evoked during periods of rapid change, when tradition is being eclipsed by new modes of thinking and living. The velocity of progress in the nineteenth century was unprecedented, and drugs became potent symbols of both its dreams and its nightmares. The notion of modernity that rose to dominance in the early twentieth century was one to which drugs represented an existential threat. Now, a century later, they are pregnant once more with possibilities for expanding our minds, restoring them to health and allowing us to discover our authentic selves.

The long sweep of modernity has been exhaustively surveyed by the Canadian philosopher Charles Taylor, whose work aims to trace the journey from the medieval worldview, in which we inhabited our allotted place in a divine scheme, to a secular modernity in which personal identity is something that we construct for ourselves. In his most sustained study, *Sources of the Self* (1989), Taylor attempted to enumerate the distinctive characteristics of the modern mind: as he defined it, 'the senses of inwardness, freedom, individuality and being embedded in nature which are at home in the modern West'.[19] This modern self, according to Taylor's analysis, sees itself as possessing inner depths that are worthy of exploration; it prizes authenticity and originality; it sees nature as benign and restorative. It is striking how closely these values reflect the attraction of consciousness-altering drugs in the twenty-first century.

As Taylor shows, however, these are not recent preoccupations: they are the culmination of a historical process that has been in motion since the Enlightenment and was firmly entrenched in popular culture during the Romantic era. It was during the eighteenth century, he argues, that Europeans came to valorise inner experience, and to analyse the mechanisms of sensation and perception that created it. During the nineteenth century this impulse broadened and deepened with the Romantic movement, which privileged 'the reflexive powers which are central to the modern subject, those which confer the different kinds of inwardness on him or her, the powers of disengaged reason, and the creative imagination'.[20] In this panoramic view, the twentieth century's taboo on the use of drugs to explore the mind appears as a hiatus in the long pursuit of our modern selves, and the current revival of interest in them as the resumption of a much older story.

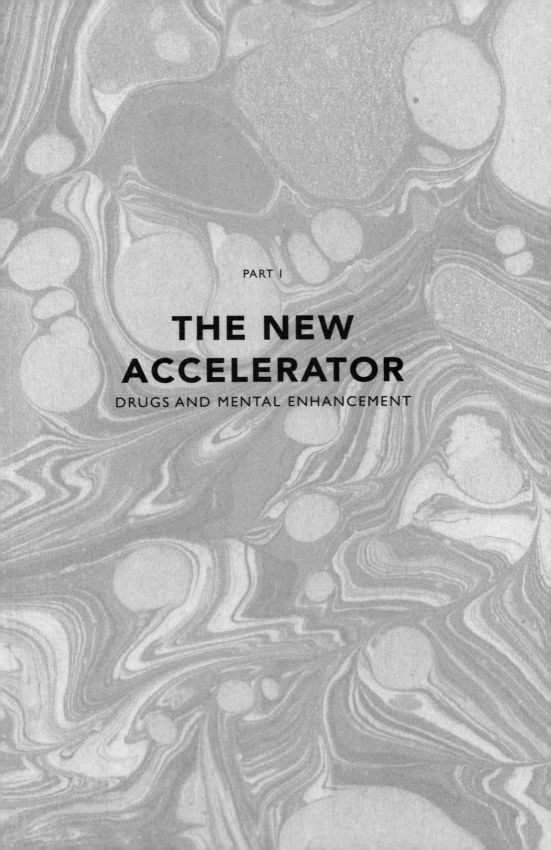

PART I

THE NEW ACCELERATOR

DRUGS AND MENTAL ENHANCEMENT

1. *Coca wine, made by steeping the cocaine-containing leaf in burgundy, was a popular product in nineteenth-century pharmacies, marketed as 'The Elixir of Life'.*

CHAPTER ONE

THE ELIXIR OF LIFE

O n 30 April 1884, Sigmund Freud received his first gram of cocaine by mail from the Merck pharmaceutical company in Darmstadt. He had been taken aback by the steep price – 3 gulden 33 kreuzer, around ten times what he had anticipated – but ordered it anyway. As he wrote to his fiancée Martha Bernays, he was only one 'lucky hit' away from a career breakthrough, and 'you know that when one perseveres, sooner or later one succeeds'.[1] In his dingy student rooms at Vienna General Hospital, he measured out 50mg of the crystalline powder, dissolved it in water and swallowed it, noting immediately that 'it has a somewhat bitter taste and produces an anaesthetic effect on the mucous membranes'.[2]

Freud's 'cocaine episode' is perhaps the best-remembered drug self-experiment of the late nineteenth century, and the most consistently misunderstood. From today's perspective, it seems baffling that the most celebrated cartographer (or *conquistador*, to use his own term)[3] of the modern mind should have staked his professional reputation on a drug that we now associate with egomaniacal excess, wild parties and Latin American criminal cartels. But in 1884, the twenty-eight-year-old Freud was merely an overworked medical student rather than the magisterial figure he would become, and cocaine was not yet the demonised drug of the twentieth century. Though his cocaine researches ended in disgrace and he subsequently disowned them, he had good reasons to hope that the drug might be a sovereign remedy for the diseases of modern life.

2. Sigmund Freud and his fiancée Martha in 1885.

He had good reasons, too, for deciding that the scientific investigation of its potential should properly begin with experiments by the investigator on himself. This was a practice that could be traced back to the beginnings of Western science, but Freud's self-experiment marked a moment at which its legitimacy was being challenged. Drugs that produced marked effects on the mind – altering moods, thoughts, feelings and perceptions – were legitimate and often highly productive subjects for scientific study, but they posed particular problems for it. The evidence that they generated was not measurable or verifiable, and could only be given from the perspective of their subject. Self-experiment had always been a practice that split researchers in two, making them both observer and subject, and forcing them to negotiate a balance between the subjective and the measurable. Freud was attempting to find this balance at a moment when the authority of the trained scientist was becoming embroiled with commercial advocacy for a booming pharmaceutical industry, and new and powerful mind-altering drugs were emerging more rapidly than the questions they raised could be answered. The drug-taking scientist was a lightning rod for these controversies, both within the fast-changing worlds of pharmacy and medicine and among an anxious but fascinated public.

* * *

Across Europe and the United States, the 1880s was a decade of velocity and acceleration. The first electric tram appeared on the streets of Berlin in 1879, and in the United States in Cleveland, Ohio in 1884. Cities were criss-crossed by cyclists, the countryside by trains, the seas by ocean liners. The telegraph system was spreading its filaments across the globe, making the price of goods in every port known instantaneously. Pocket watches were sold by the millions as working people needed to synchronise their plans, coordinate with travel timetables and manage their schedules in precise increments. This was a new way of living, and there were many who believed it was placing a stress on the brain and nervous system for which humanity was unprepared and ill-adapted. New diagnostic terms such as 'railway spine' and 'bicycle face' were coined for particular syndromes attributed to movement at speed, while the pervasive stresses of modern life were captured in the diagnosis of 'neurasthenia', a term popularised in 1869 by the American neurologist and physician George Miller Beard.

Beard was a powerfully focused and energetic character with an unlimited confidence in the explanatory power of science. He did not invent the term, but he was the first to describe neurasthenia as a distinct organic disease, and to diagnose its cause as modernity itself. He described its biological mechanisms in precise detail. It had been recognised since the eighteenth century, when Luigi Galvani demonstrated that the legs of dissected frogs could be made to twitch by passing an electric current through them, that the nervous system was driven by electrical forces. As electrical theory developed, it encouraged a parallel model of anatomy in which the brain was a voltaic battery that regulated the body's electrical force, and the nerves were the electrical wires that transmitted it. From this emerged the idea of a 'nervous economy': if nervous energy was spent faster than it accumulated, the inevitable consequence would be 'nervous exhaustion' and the collapse of the system.[4]

The symptoms of neurasthenia could vary widely, but they commonly included anxiety, dizziness, headaches, indigestion, nervous spells, chronic fatigue, brain fog, insomnia and impotence. None of these was new, but in his bestselling book, *American Nervousness* (1881), Beard pointed repeatedly to the reason why they had reached epidemic proportions: 'First of all, modern civilisation.'[5] The pace of life and the demands of business and industry were pushing the human machine beyond its natural limits. The ubiquity of pocket watches was both a symptom and a cause of the disease:

> The perfection of clocks and the invention of watches have something to do with modern nervousness, since they compel us to be on time, and excite the habit of looking to see the exact moment,

so as not to be late for trains or appointments . . . A nervous man cannot take out his watch and look at it when the time for an appointment or train is near, without affecting his pulse, and the effect on that pulse, if we could but measure and weigh it, would be found to be correlated to a loss to the nervous system.[6]

Modern technology provided Beard with a topical metaphor for the process. 'Edison's electric light', he wrote, 'is now sufficiently advanced in an experimental direction to give us the best possible illustration of the effects of modern civilisation on the nervous system.'[7] The human metabolism, like an electrical circuit, contains finite reserves of energy. When the circuit is more and more heavily loaded, there comes a point when 'the amount of force is insufficient to keep all the lamps actively burning . . . this is the philosophy of modern nervousness'.[8] The accelerating diffusion of print and information, the rise in intensive education, and the scientific advances that were remaking daily life: 'all these are so many additional lamps interposed in the circuit, and are supplied at the expense of the nervous system, the dynamic power of which has not correspondingly increased'.[9]

Neurasthenia became the disease of the age, not only through the efforts of Beard and his fellow nerve specialists, but by public demand. Most diagnoses of mental illness carried a powerful stigma of insanity or moral weakness. The diagnosis of neurasthenia, by contrast, implied a sense of martyrdom or heroism: a build-up of external stress that the patient had fought valiantly to the point of exhaustion, when their nerves had given way. Its symptoms were extremely common and, unlike other common diseases – for example tuberculosis or cholera, for which the German physician Robert Koch had recently identified microbial causes – there was no biological marker or medical test to confirm or reject the diagnosis. Beard himself treated it with electrodes that passed a mild current across the body, a method that achieved some success, though he suspected that part of this was 'mental therapeutics', or what we would now call the placebo effect. The only real remedy, he believed, was a root-and-branch transformation of society to reduce the intolerable stresses heaped on its citizens and foster the emergence of a less neurotic and healthier nation.

Sigmund Freud diagnosed himself as neurasthenic, and attributed his condition to a career path that had left him overworked, stressed, exhausted and often depressed. It had taken him eight years to complete his medical doctorate rather than the usual five because he had simultaneously been doing laboratory research on invertebrate nerve fibres: he was pushing forward on twin tracks, one towards a private practice as a neurologist and the other as a university specialist. His initial scoping of the medical literature on cocaine – which at this point amounted mostly to studies of its natural source, the leaves of the Andean coca bush – suggested that it had been 'almost universally effective in improving those functional disorders which we now group together under the name of neurasthenia'.[10] As he proceeded with his researches and experiments, he became convinced that the drug might be able to supply precisely what Beard believed the human nervous system lacked: a means of increasing its capacity, rather than merely giving a temporary boost to its energies at the price of subsequent depletion and exhaustion. If all the varied symptoms of neurasthenia sprang from the same cause – a deficit of nervous energy – the right drug might relieve them all.

'It is a well-known fact', Freud observed in his first published report on the subject, 'Über Coca' (1884), 'that psychiatrists have an ample supply of drugs at their disposal for reducing the excitation of nerve centres, but none which could serve to increase the reduced functioning of the nerve centres.'[11] The nineteenth-century pharmacopoeia was by this point rich in narcotics and sedatives: opium, the old staple, had been joined by its pure and concentrated extract, morphine, as well as anodyne vapours such as ether and chloroform, and the powerful sedative compound chloral hydrate. Combinations of these drugs in patent medicines, prescribed by physicians and sold over the pharmacy counter, were widely used to treat neurasthenic symptoms such as exhaustion, anxiety and overwork. But these were only palliatives that allowed patients a temporary respite before the unequal contest between modern civilisation and the human nervous system resumed. A stimulant that changed the equation by adding energy to the system might offer a genuine cure.

Freud was, naturally, not the first person to have investigated such a possibility. Several drugs had been pressed into service as stimulants, but each had its limits. The most readily available was alcohol, which was widely used in medicine, including as a pick-me-up in cases of exhaustion, injury and weakness. But the boost in energy that it offered was short-lived, and was succeeded by a depression of the metabolic system that often caused a relapse into debility and stupor.

Unlike other regions of the globe, which were home to caffeine-containing plants as well as other stimulants such as coca, khat or betel, the flora of western Europe had no native energy boosters, and when tea, coffee and chocolate arrived during the seventeenth century they were hailed as medical marvels. They were, in the paradoxical phrase of the time, 'sober intoxicants' that could revive the sick and energise the healthy, increase stamina, sharpen the mind and banish sleep. In large quantities, however, they overstimulated, causing unpleasant side-effects such as nervous tremors, sweating and a racing heartbeat, overstraining the nervous system rather than boosting it.

These high-dose effects were described by the author and prodigious coffee drinker Honoré de Balzac, who drank dozens of cups a day while writing his novels and, in his 1839 *Treatise on Modern Stimulants*, admitted that 'I consume it in such quantities that I have been able to observe its effects on an epic scale'.[12] Balzac's coffee consumption mounted to the point where 'finally, I discovered a terrible and cruel method, which I would recommend only to men of excessive strength':[13] swallowing handfuls of ground coffee beans without water on an empty stomach until:

everything becomes agitated: ideas march like the battalions of a great army onto the battlefield where the battle has begun. Memories charge in, flags flying; the light cavalry of comparisons advances at a magnificent gallop; the artillery of logic rushes in with its convoy and its charges; witticisms appear like snipers; characters rise up; the paper covers itself in ink, because the evening begins and ends with a torrent of black water, as the battle does with its gunpowder.[14]

Balzac managed astonishing feats of literary production on these doses, but it was a Pyrrhic victory: an assault on the stomach and the brain that brought 'terrible sweating' and 'a weakened nervous system' that led him, eventually, to abandon his experiments. The effects of caffeine at this pitch of intensity were frantic and unstable, too easily dissipated in impatience, frustration and anger. Like alcohol, too, it reclaimed its temporary loan of energy with interest: it was impossible the following day to ignore the realisation that 'coffee wanted a victim'.[15]

During the 1850s, a range of 'nerve tonics' had appeared in pharmacies, but all were toxic and self-limiting. Strychnine was promoted as a remedy, in three small doses a day: in 1884, just as Freud was embarking on his experiments, the powerhouse of British science T.H. Huxley had added it to his regime for treating exhaustion ('I suppose that everybody starts life with a certain capital of life stuff & that expensive habits have reduced mine').[16] Arsenic was another chemical tonic, the active ingredient in common patent medicines such as Fowler's Solution: ten drops, three or four times a day, could restore the nerves, but it could equally bring on neurasthenic symptoms such as nausea, fatigue and headaches. Calomel – mercuric oxide, dispensed in the form of 'blue pills' – was recommended for boosting nervous energy as well as treating inflammation and stiffness, but this too was toxic in anything other than small doses.

Neurasthenia treatments also encompassed a thriving market in electrical devices that claimed to boost the system by passing current through the body, and by Freud's time medical researchers were exploring drug therapies far more alarming than cocaine. Charles-Edouard Brown-Séquard, a professor of medicine at the Collège de France in Paris, had

3. Cocaine arrived in a marketplace where poisons such as strychnine, arsenic and mercury were sold as nervous stimulants.

been experimenting for some years with glandular extracts, particularly from animal testes, on the basis of his belief that life-energies were concentrated in semen and the glands that produced it. 'If it were possible', he speculated in 1869, 'to inject semen into the blood of old men, we should probably obtain manifestations of increased activity as regards the mental and the various physical powers.'[17] In 1889, at the age of seventy-two, Brown-Séquard announced to members of the Société de Biologie in Paris that he had made a preparation of semen, blood and crushed testicles from dogs and guinea pigs and given himself a course of ten subcutaneous injections. He reported a return of youthful vigour and increased 'facility of intellectual labour', though it seems implausible that his serum could have been biologically active.[18] Nonetheless, his 'organotherapy' became fashionable across Europe and America, and was touted by therapists not merely as a remedy for neurasthenia but – in a phrase also used by marketeers of coca and cocaine products – as 'the elixir of life'.

When set against the poisons sold over pharmacy counters, electropathic corsets or Brown-Séquard's self-experiments, Freud's investigation of the stimulant powers of cocaine seems not reckless but positively sober. Historians and biographers, as well as the older Freud himself, have been quick to cite youthful ambition as a primary motive for his research; but if he had an impetuous streak, he also had a cautious one. On becoming engaged to Martha Bernays, he asked her to embroider for him two samplers, or 'votive tablets', to brighten up the walls of his drab university rooms. One was to read 'Travailler sans raisonner' ('Work without reasoning') – a recipe for neurasthenia – and the other 'En cas de doute abstiens-toi': 'When in doubt, abstain'.[19] His engagement to Martha, the daughter of a Jewish family wealthier and of higher status than his own, was a romantic match that he celebrated in high literary style in their correspondence, but it was also bound up with his keen hunger for success; and if cocaine could make his fortune, it would smooth his path to an advantageous marriage. But his urge for discovery coexisted with his injunction to self-control. His self-experiments came to be seen as impetuous and excessive, but with hindsight their more serious flaw was his reluctance to test the new drug to its limits.

Freud's first letter to Martha on the subject of cocaine records his initial steps. His attention had first been drawn to the drug by a report in a medical journal, the *Deutsche Medicinische Wochenschrift*, the previous year. An army physician, Theodor Aschenbrandt, had secretly added cocaine to the drinks of his Bavarian recruits during their annual weapons drill and observed that it made them better able to endure hunger, heat, strenuous exercise and marches. 'In this way,' he concluded confidently, 'the soldier can dispense with food for eight days.'[20] Cocaine, Freud learned, had been isolated in 1860 from the leaves of the coca plant, after which it was named; since then it had been available from Merck's pharmaceutical catalogue, but little investigated. Its high price was a deterrent, particularly since the scattered assessments of the coca leaf's potency were inconclusive. Many of the samples shipped from South America to Europe over the years lost their stimulant properties on the journey, and some chemists had concluded that their reputed effects were based on exaggerated reports from the natives of the Andes.

In the USA, however, cocaine was being enthusiastically promoted, particularly in the *Therapeutic Gazette*, a monthly journal published in Detroit under the aegis of the Parke, Davis pharmaceutical company. Parke, Davis had built up an unsurpassed range of medical products, many extracted from plant sources: their bestseller, the laxative cascara, was derived from a bark long used medicinally by the indigenous peoples of the Pacific north-west. In 1880 the *Therapeutic Gazette* published a glowing testimonial to cocaine by the physician W.H. Bentley, who recommended it for the treatment of opium, morphine and alcohol habits. Cocaine, Bentley wrote, was 'capable of producing the most exalted mental feelings, far more ecstatic than anything ever experienced from the use of opium or alcohol'; and unlike these other drugs, 'its effects pass away gradually after a few hours, leaving a feeling of buoyant serenity, not to be succeeded by any depression'.[21]

Bentley appended a series of case histories: as was usually the case in journals tied to pharmaceutical companies, all were successful cures. In 1884 Parke, Davis added a tincture of coca to their catalogue and dispatched their most experienced botanical researcher, Henry Hurd Rusby, to Bolivia with instructions to secure a large supply of coca leaves. Rusby brought back a huge haul of 200,000 leaves but discovered, as others had before him, that most had spoiled on the return journey; he proceeded to develop a technique for preserving the cocaine yield by making a crude but stable extract in makeshift laboratories in South America. With their new supply of pharmaceutical cocaine in the pipeline, Parke, Davis were poised to become the world's largest supplier, and to bring its exorbitant price down to earth.

Aschenbrandt's paper directed Freud to the extant literature on the coca leaf in Spanish, German and Italian. Much of it derived from self-experiments, and had previously been collected by the aristocratic explorer and biochemist Baron Ernst von Bibra in his 1855 monograph, *Plant Intoxicants*. Von Bibra travelled widely in South America before settling into domestic life in his family castle in Bavaria, surrounded by his extensive collection of art and architecture, where he installed a large home laboratory in which he analysed the contents of plants, foods and drugs. His account of coca quoted extensively from the botanist Eduard Pöppig, who had travelled in the Andes in the 1830s, and presented the leaf not as a beneficial stimulant but the deleterious habit of a primitive race. 'The lower the mental faculties of a nation,' Pöppig wrote, 'the coarser the narcotics it derives pleasure from.' In his account, coca's habitual users became pale, weak and sick, causing them to waste away, and yet 'it has never been possible to break the vice of a *coquero*, as a true coca addict is called in Peru'.[22]

Von Bibra's personal experience was quite different: 'Pöppig's judgment seems to us rather hard and severe, in regard to both the psychological and physiological effects.'[23] During his own visit, he had several opportunities to 'get acquainted with the pleasures of a true *coquero*', and to master the delicate technique of chewing the leaf in combination with the slaked lime powder that releases its alkaloids. He had observed its use as an aid to energy and productivity, chewed by agricultural workers and particularly by those undertaking long journeys on foot. However, von Bibra had personally found its stimulant effects to be modest: apart from a mild loss of appetite, he noted 'no sensation in me that might have indicated a nervous excitement'.[24]

A far more extensive report, and the one that influenced Freud most powerfully, was that of the Italian neurologist Paolo Mantegazza, who practised as a doctor in Argentina and Paraguay in the 1850s and self-experimented with the local plant stimulants, guarana and coca. Unlike von Bibra, Mantegazza appreciated coca's stimulant effects immediately and pursued them vigorously. 'As soon as one chews one or two drachms',[25] he wrote in his 1859 monograph 'On the Hygienic and Medical Values of Coca',

> the nervous excitement is always followed by movements that are exaggerated or violent, and always irregular; there is a general confusion of thoughts and muscular activity, while in the inebriety produced by coca it seems that the new strength gradually drenches one's organism in every sense, as a sponge soaks itself with water. Thus the delight of the period consists almost completely in an increased consciousness of being alive.[26]

Mantegazza found in coca not a productive stimulant for the sober self, but a radically altered state of consciousness. Unlike caffeine, higher doses brought not overstimulation but ever more pleasurable and remarkable effects. By chewing 8 drachms in a day and a further 10 the same evening, about the most he could physically manage, he attained what he called 'the delirium of coca intoxication, and I must confess that I found this pleasure by far superior to all other physical sensations previously known to me'.[27] He recorded his pulse before the evening dose at 83 per minute; half an hour later it had risen to 120. He felt supremely happy, and on closing his eyes was presented with 'the most splendid and unexpected phantasmagoria', kaleidoscopic images succeeding each other too fast to record, or even to communicate by announcing them in rapid fire to the colleague beside him. He attempted to transcribe them, missing ten for every one he managed to capture:

> A cave of lace through the entrance to which can be seen, toward the back, a golden tortoise seated on a throne made of soap . . .

A battalion of steel pens fighting against an army of cork-
screws . . .

Lightning, consisting of glass threads, piercing a whole Parmesan
cheese crowned with ivy and berries . . .

A saffron inkwell from which is born an emerald mushroom
studded with rose fruits . . .

A ladder made of blotting paper lined with rattlesnakes from
which several red rabbits with green ears come jumping down . . .[28]

Mantegazza embraced coca's euphoric and visionary properties, which convinced him that 'all this will be great science in the near future'.[29] The desire and capacity for ecstasy was a constant throughout human history, but he believed that its limits were still unexplored. On his return to Italy he became a medical professor in Pavia, founded the Italian Anthropological Society and began work on a massive survey of inebriation and human nature, which ran to 1,200 pages when it was eventually published in 1871. His expansive category of inebriants included non-chemical stimuli such as joy, love, ambition, youth and religious ecstasy: 'each passion', he argued, 'may have a paroxysm resembling inebriation'.[30] He placed the pleasure produced by drugs such as coca at the heart of his vision of a progressive future in which humanity would attain a new form of 'inebriation caused by far-sightedness and optimism'. As science advanced, nature and chemistry would reveal 'a thousand new nervine nourishments', and a society would arise in which 'men will always have feasts and inebriations'.[31] For his part, Mantegazza continued to use cocaine in a 'wise and copious' manner into his advanced old age.[32]

In his first paper on the topic in 1884, 'Über Coca', Freud adopted a strikingly innovative literary style, quite different from anything that he would write subsequently, which aimed to balance the sober, quantitative approach of Aschenbrandt's researches with the intoxicating qualities of Mantegazza's. He began with a conventionally written botany and history of the coca plant, but in the physiological section, 'The Effects of

Coca on the Healthy Human Body', he switched abruptly to first-person testimony: 'I have carried out experiments and studied, in myself and others, the effects of coca on the healthy human body; my findings agree fundamentally with Mantegazza's description of the effects of coca leaves.'[33] He described his first dose with clinical precision – 0.05g of cocaine in a 1 per cent water solution – but then shifted voice once more, to the indirect conversational form, 'you' or 'one':

A few minutes after taking cocaine, one experiences a sudden exhil-aration and feeling of lightness. One feels a certain furriness on the lips and palate, followed by a feeling of warmth in the same areas; if one now drinks cold water, it feels warm on the lips and cold in the throat.[34]

As he moved on from the physical to the psychological effects, the voice shifted yet again, now alternating the personal address to the reader with the impersonal tone of scientific authority:

The psychic effect of *cocaïnum muraticum* in doses of 0.05–0.10g consists of exhilaration and lasting euphoria, which does not differ in any way from the normal euphoria of a healthy person. The feeling of excitement which accompanies stimulus by alcohol is completely lacking; the characteristic urge for immediate activity which alcohol produces is also absent. One senses an increase of self-control and feels more vigorous and capable of work; on the other hand, if one works, one misses that heightening of the mental powers which alcohol, tea or coffee produce. One is simply normal, and soon finds it difficult to believe that one is under the influence of any drug at all.[35]

At times Freud describes cocaine with a breathless energy, intended to convey the subjective sensations produced by the drug and inviting the reader to share in the frank acknowledgement of its pleasures. At others, he portrays it as a 'sober intoxicant' with no significant subjective effects.

In some passages he strikes a tone of objectivity and professional distance; others are styled as personal confession, in which he claims the authority of experience. The conversational 'one' or 'you' occupies a slippery middle ground between the two: intimate and direct but not, on close inspection, based firmly on either evidence or experience.

These shifts in style were a carefully modulated solution to the paradoxes of self-experiment, which were well recognised. Ernst von Bibra, similarly, had commented in his essay on coca that 'two persons, as it were, seem to be present, one feeling all the effects of the narcosis, the other conscious of them'.[36] He might have added a third perspective: that of the author, balancing the voices of observer and subject to present the experience to the reader. Looking back on his cocaine researches, Freud would be more specific: 'I realise that such self-observations have the shortcoming, for the person engaged in conducting them, of claiming two sorts of objectivity', the perspective of the researcher and that of the experimental subject.[37] But these shifts in style also reflected Freud's ambition to address several different audiences at once: his medical colleagues and superiors, the pharmaceutical trade on which cocaine's success or failure would depend, and the general public who would be its eventual customers. He was attempting to present himself simultaneously as a medical expert, a salesman for a revolutionary new pharmaceutical, and a *conquistador* of the mind. In doing so he needed to unite two opposing schools of the self-experimental method that had coexisted from the origins of modern science: one demanding professional distance from the subject of his inquiries, the other pushing him deeper into the terrain of introspection and subjectivity.

* * *

'Nullius in Verba': the motto of the Royal Society, adopted on its foundation in 1660, placed experiment at the heart of the scientific revolution. It announced that nothing was to be taken on hearsay: direct evidence, generated by experiment and first-person observation, was to replace the traditional reliance on classical and scholastic authorities. Experiments, in turn, were to be confirmed by replication. When they were demonstrated in the semi-public space of the Royal Society, a register book was supplied for witnesses to testify and endorse what they had seen. The readers of the Society's published *Proceedings* were invited to repeat the experiments themselves, and to write in with evidence that confirmed or denied them.

The Royal Society's leading figures, including Robert Boyle and Isaac Newton, developed mechanical and material theories that created a distinction between two different types of evidence. 'Primary' qualities such as size, shape or weight were directly measurable and, as we would say today, objective; 'secondary' qualities were those such as texture, taste or feeling that described human sensations and responses. The Society's researches favoured the demonstration of primary qualities, but sensation and perception were legitimate fields of inquiry, and certain classes of data could only be demonstrated by self-experiment. In a famous and graphic example, when Isaac Newton wished to establish whether a change in the curvature of the eye would present a distorted image to the perceiver, he took a large needle, or bodkin, and:

> put it betwixt my eye & the bone as neare to the Backside of my eye as I could: & pressing my eye with the end of it (soe as to make the curvature a, b, c, d, e, f in my eye) there appeared several white darke & coloured circles r, s, t, &c. Which circles were plainest when I continued to rub my eye with the point of the bodkin, but if I held my eye & the bodkin still, though I continued to presse my eye with it yet the circles would grow faint & often disappeare untill I renewed them by moving my eye or the bodkin.[38]

The letters are keys to Newton's accompanying drawing, which shows the shape of the eye indented. It was not possible to present direct evidence for the coloured circles since they only existed in Newton's mind, yet his fellow philosophers were not obliged simply to accept

Newton's word for them. They were described in such a way that any sceptic, if they so wished, could repeat the experiment on themselves.

This was also the case for drugs that acted on the mind. Like the circles in Newton's vision, the changes in thought, mood, sensation or perception they produced were secondary qualities with no material existence, but this did not mean they were delusions. There was no correct or perfect way to present them, yet they offered unique insights into mental functioning. The language that offered itself most plausibly for describing them was that of medicine, in which the physician could only report the patient's sensations and state of mind at second hand, but could add judicious glosses and interpretations suggested by their professional learning and experience of similar cases.

An early example was Robert Hooke, curator of experiments at the Royal Society, whose diary records the effects of the many drugs he used for overlapping purposes: verifying medical claims, managing his pain and moods, and 'refreshing' himself during social and business meetings in the coffee houses he frequented in the afternoons, following his morning routine of experiments and instrument-making.[39] He recorded his impressions of alcohol, chocolate, tea, coffee and tobacco, and on 18 December 1689 delivered a lecture at the Royal Society entitled 'An Account of the Plant, Call'd Bengue', or cannabis. 'It is a certain Plant which grows very common in India', Hooke related, 'and the Use thereof (though the Effects are very strange, and at the first hearing frightful enough) is very general and frequent':

> This Powder being chewed and swallowed, or washed down, by a small Cup of Water, doth, in a short Time, quite take away the Memory & Understanding; so that the Patient understands not, nor remembereth any Thing that he seeth, heareth, or doth, in that Extasie, but becomes, as it were, a mere Natural, being unable to speak a Word of Sense; yet is he very merry, and laughs, and sings, and speaks Words without any Coherence, not knowing what he saith or doth; yet is he not giddy, or drunk, but walks and dances

and sheweth many odd Tricks; after a little Time he falls asleep, and
sleepeth very soundly and quietly; and when he wakes, he finds
himself mightily refresh'd, and exceeding hungry . . .[40]

Hooke begins by announcing that he has a fantastical-sounding tale
to report, and takes care to set it in its geographical context, but his
impersonal narrative voice leaves the provenance of the story in doubt.
Who was the patient or subject, and who was the observer? The author
identifies himself as neither, though he may have been either or indeed
both. Hooke's usual inclination was to experiment on himself, and in
his diaries he writes of drugs in the first person ('I took spirit of urine
and laudanum with milk for three preceding nights. Slept pretty well').[41]
Yet in this case the observer may equally have been his friend Robert
Knox, a sailor who had passed on a sample of the plant extract to him
when they met in a London coffee house in November 1672. Knox had
a remarkable story: he had been held captive in Sri Lanka for several
years before escaping on a stolen ship and almost dying of fever, from
which he was rescued only by his discovery that cannabis worked as 'an
Antidote and Counter-Poyson against the filthy and venomous water'.[42]

By the new standard of 'Nullius in Verba', the testimony Hooke
presented fell somewhat short of scientific evidence. Travellers' tales of
exotic drugs and miracle cures were an inheritance from classical literature, where they inhabited a murky borderland between history, botany
and folklore, much as they still did in the seventeenth century. Hooke
was offering the Society a plausible account, but one that was not subject
to checking except by self-experiment. What emerged was something
like a medical case history: anecdote rather than data, but mediated and
endorsed by an expert witness. Individual and bodily experience – of the
kind elicited by doctors with the question 'how do you feel?' – was never
able to produce the kind of replicable results generated by the Society's
vacuum pumps or thermometers. In its place an informal criterion of
expertise emerged, dubbed by the historian of science Simon Schaffer
'the Cartesianism of the genteel': the assumption that trained or educated

observers were capable of using their minds to assess the evidence of their bodies – or, in the terms established by the philosophy of John Locke, of separating the intellect from the passions.[43] As in Hooke's case, it became the convention to relay such evidence in the third person or passive voice favoured by physicians, with the confessional first person confined to diaries and private records.

As pharmacology developed through the eighteenth century, this language of quasi-medical reportage became firmly entrenched. Self-experiment with drugs was common practice, for ethical as well as practical reasons: physicians were obligated to treat the sick according to the Hippocratic oath, doing no harm, and experimenting on them was the mark of an unscrupulous quack. Working with medicines of unknown purity, and with only a partial grasp of their active chemical principles, doctors reported experiments as far as possible through external measurement of physical signs: pulse, temperature, stools or urine. Practising physicians were aware that responses to drugs could vary widely between individuals, and prescribed individually tailored regimes of medicines and other treatments to their fee-paying clients. Administering drugs was the business of apothecaries, a lower status profession whose members were more likely to deal in standard doses.

The growth of physical observation and measurement revealed that mind and body influenced each other in mysterious ways, and that even directly experienced phenomena could be proven false. One famous example was the committee set up in 1784 by the French Royal Academies of Science and Medicine, under the aegis of Benjamin Franklin, to investigate Anton Mesmer's theory of animal magnetism or mesmerism. A series of blinded experiments, in which magnetised and unmagnetised objects were swapped without the subjects' knowledge, showed that the dramatic effects of 'animal magnetism', from paralysis to contortions to spontaneous healing, could manifest even when no magnetism had been applied. This effect, which we would now call placebo, even affected some members of the committee, who felt twinges and electric sensations, though to a lesser degree than mesmerism's true believers. 'There is no proof of the existence of the Animal Magnetic fluid,' the committee's report concluded; 'the Imagination without the aid of Magnetism can produce convulsions, but Magnetism without the Imagination can produce nothing.'[44] The lesson for self-experimenters was that hypervigilance to one's own sensations could overinterpret the evidence of the body or generate symptoms based solely on expectation. As the final report stated, 'the first thing therefore, to which the commissioners were bound to attend, was not to observe too minutely what passed within them'.[45]

For the generation that followed, however, subjectivity was the new frontier of scientific knowledge. The inward turn emerged from the philosophy of Immanuel Kant, whose treatise of 1781, *Critique of Pure Reason*, made a primary distinction between the 'phenomenal' world – reality as revealed by sensation and perception – and a 'noumenal' world of ideas and categories, including God, that existed prior to and independently of human experience. According to Kant's distinction, the world as received via the senses was not the accurate reflection of an external reality but a construct, shaped by the human senses and limited by the parameters of the human mind.

This theory was strikingly corroborated in a series of drug experiments by the young chemist Humphry Davy that were still commonly cited as a touchstone for the practice of self-experiment in Sigmund Freud's day. In 1799, at the age of twenty, Davy was hired as the chemical assistant at the Medical Pneumatic Institution in Bristol, an experimental project initiated by the pioneering physician Thomas Beddoes for synthesising and testing gases in the treatment of lung conditions. One of the first compounds Davy created in the laboratory was nitrous oxide, a recently discovered gas that was believed to be highly toxic. Davy suspected this belief resulted from a confusion with a related compound, nitric oxide, a red-brown gas that was a powerful irritant. 'I made a discovery yesterday which proves how necessary it is to repeat experiments,' he wrote to his and Beddoes's friend Davies Giddy in April 1799; 'The gaseous oxide of azote [nitrous oxide] is perfectly respirable when pure.'[46]

Excited to have established a virgin field of inquiry, Davy and Beddoes heated ammonium nitrate crystals in an alembic and collected the escaping gas in an air holder, from which Davy inhaled through a breathing tube. As he filled his lungs, he noticed an unexpected sensation, 'a highly pleasurable thrilling in the chest and extremities'. As he continued, 'the objects around me became dazzling and my hearing more acute', and the sensations built towards a climax in which 'the sense of muscular power became greater, and at last an irresistible propensity to action was indulged in'. Beddoes recorded that Davy leapt violently around the laboratory shouting for joy. For his own part, Davy retained only vague recollections of these ecstatic moments, and were it not for the scrawled notes he discovered the following morning, 'I should even have doubted their reality'.[47]

Unlike the speculative animal fluids of Dr Mesmer, there was no doubt about the material cause of this paroxysm of pleasure. Nitrous oxide – first isolated by Joseph Priestley, who had named it 'dephlogisticated nitrous air' – was a chemical substance with a known synthesis, and the experiment could be replicated and verified in any home laboratory. The effects of the gas, however, could only be captured in first-person testimony. Davy was quick to elicit such testimony, and was perfectly placed to do so. The Pneumatic Institution was a hub for Bristol's freethinking writers, philosophers and physicians, and over the summer of 1799 dozens of them came to visit and experience the gas which the poet Robert Southey, the first of his friends to whom Davy had offered it, described as 'the wonder-working air of delight'. After his first dose, Southey wrote to his brother that 'Davy has actually invented a new pleasure for which language has no name'.[48] Exploring it further was an invitation and a challenge that the keenest minds of Bristol were eager to accept.

As the experiments progressed, Davy realised that a new 'language of feeling', as he called it, was required to describe the effects of the gas. The standard question of medical description, 'How do you feel?', was tested to its limits by a torrent of sensations that encompassed dizziness, tingling, a sense of mental exhilaration and onrushing cosmic epiphany that rapidly dissolved into incoherence and, frequently, hysterical laughter with no obvious cause. Davy and Beddoes attempted a few trials on patients with lung diseases; one responded to the question with, 'I do not know, but very queer.' Another responded, obliquely but suggestively, 'I feel like the sound of a harp.'[49] During the evenings they experimented further on healthy volunteer subjects. When Davy set the chemical reaction bubbling and offered a new subject a green silk bag of the gas, he often began by giving them a dose of ordinary air to rule out any elements of suggestion or expectation. Once they had recovered from a lungful of the real thing, he asked them to write a brief description of their experience. As one of the volunteers, the surgeon Thomas Hammick, wrote after his intoxication, 'We must either invent new terms to express these new and peculiar sensations, or attach new ideas to old ones, before we can communicate intelligibly with one another on the operations of this extraordinary gas.'[50]

Nitrous oxide, for Davy and his circle, collapsed the distinction between the intellect and the passions: it stimulated both, with equal intensity. It was a profoundly embodied experience, susceptible to external measurements – for example, the amount of gas that was inhaled, or had dissolved into the bloodstream – but not reducible to them. It asked profound questions about the relation between mind and body: how could inhaling an artificially created chemical affect not merely the breathing and the pulse, but the emotions, the sense of wonder and the imagination? Davy's ground-breaking report on the experiments, *Researches Chemical and Philosophical, Chiefly Concerning Nitrous Oxide and its Respiration* (1800), yoked body and mind, intellect and passion together with a structure that ascended from the chemical to the medical to the sublime. Its opening section was a description of nitrous oxide's chemistry and synthesis; the next was a precise account of its physiological action, and it concluded with the subjective reports of over thirty volunteer subjects. Davy's own contribution, after absorbing as much gas as humanly possible by enclosing himself in an airtight box filled with it for an hour and a quarter, gave the language of feeling full rein:

> I heard every distinct sound in the room and was perfectly aware of my situation. By degrees as the pleasurable sensations increased, I lost all touch with external things; trains of vivid and visible images rapidly passed through my mind and were connected with words in such a manner, as to produce perceptions perfectly novel. I existed in a world of newly connected and modified ideas. I theorised; I imagined I made discoveries . . . As I recovered my former state of mind, I felt an inclination to communicate the discoveries I had made during the experiment. I endeavoured to recall the ideas, they were feeble and indistinct; one collection of terms, however, presented itself. . . . *Nothing exists but thoughts! The universe is composed of impressions, ideas, pleasure and pains!*[51]

Davy's scientific ambition demanded that he push his experiments to the limit and experience the effects of the drug at their most intense. This in turn required a break with the impersonal conventions of medical reportage in favour of a first-person testimony that fused the roles of observer and experimental subject. He developed his 'language of feeling' in parallel with that of the young poets among his volunteers, Robert Southey and Samuel Taylor Coleridge, who were also seeking a novel and introspective language to capture feelings and states of mind never previously described. Davy aspired to be a hero of science, comparing himself in his youthful notebooks to Sir Isaac Newton, but his version of science took on the qualities of the dawning Romantic age and of its most exalted quality, genius. In theory, there was no role for genius in experimental science, since data was replicable and detachable from individual personality; but self-experiment, in Davy's hands, had produced results inseparable from the dazzling mind of its subject.

This approach to knowledge found a home in the new German sciences, for example in Johann von Goethe's studies of the subjective qualities of colour perception, and in particular the emerging theories of *Naturphilosophie* that aimed to penetrate beneath the visible surface of the material world and to grasp the invisible currents and fluxes that generated it. Human consciousness and the natural world, its proponents argued, were mirrors of one another, and the observer was an integral part of the phenomena they were investigating. The principle was vividly demonstrated by electrical researchers such as Alexander von Humboldt and the flamboyant young chemist Johann Wilhelm Ritter, who moved from testing current on animals and volunteers to turning it on themselves. Ritter was, like Davy, an early experimenter with the voltaic pile, a chain of battery cells arranged to deliver a powerful standing charge, and he progressed from making the legs of dissected frogs twitch to closing a circuit of 100 batteries with his own body. He recorded his sensations when current was applied to different organs, and found he could generate colours and sounds by discharging it into his eyes and ears. 'Since one can only perform them on oneself', Ritter noted, 'these experiments are somewhat painful'; yet he persisted until he achieved sublime effects:

> I abide with my eye on the positive pole of a rather powerful column of 100, 150, or 200 layers, thoroughly lubricate the hand that closes the circuit with saline or ammonium chloride solution, sheath it well with metal and now I close, first with only a few layers, then with progressively more, until finally the whole column is in the circuit. In the beginning I have the same blue as usual; it grows stronger the further I go; but finally it stands still, darkens, a mixed colour of a greenish sort emerges, though not so distinctly green as the previous light was blue; then it develops into yellow, etc., until finally it is the most glorious red and of an intensity that I have never seen before, even on the negative pole.[52]

Like Newton's self-experiments with his bodkin, subjective visions of this kind could in theory be confirmed by other researchers; but

Davy and Ritter represented a new breed of experimenter, both rigorous mechanic and inspired genius, prepared to take scientific discovery to heroic limits. As a young physician in thrall to the Romantic generation of Goethe, Humboldt and Ritter, Freud inherited the conviction that science required more than diligent physiological observation of the kind that had thus far defined his career. After several years spent dissecting the nerves and spinal cords of lampreys and crayfish, 'Über Coca' was his bid not only for professional success, but to write himself and his experience into his work.

*   *   *

The early nineteenth century was also a golden age of drug discovery. In 1803, four years after Davy's nitrous oxide experiments, a young German pharmacy apprentice named Friedrich Sertürner began experimenting with a tarry opium concentrate, attempting to reduce it to its acidic components. Like others, he had noticed that samples of opium differed in potency and suspected that the gummy resin must contain active compounds in varying concentrations. Sertürner's researches consumed many years and produced several mysterious substances, which he tested on himself and others. Eventually, in 1817, he isolated a compound that formed clear crystals soluble in acid, though only slightly so in water. He enlisted three neighbours, teenage boys, to drink a solution of the crystals with him in cautious half-grain increments, but the drug was far stronger than he had anticipated. He and his fellow subjects suffered violent vomiting fits and fell into a heavy stupor, from which they were only revived by drinking strong vinegar. Sertürner christened his extract morphine, after Morpheus, the ancient god of sleep.

This was the first time that a plant had given up not merely an essence or a tincture, but a pure and previously unknown chemical substance. In the process, Sertürner placed self-experiment at the centre of the procedure and made several observations that laid the groundwork for isolating further pharmaceutical plant extracts. Having painfully established over a decade that his active components were not acids, as he and others had expected, he classed them as 'vegetable alkali', later to be known as alkaloids. He also established that crystallisation was the best way to isolate these chemicals and, once isolated, to keep them free from contamination by other substances. These precepts informed the isolation of caffeine in 1819 – by a young chemist, Friedlieb Ferdinand Runge, as a result of an encounter in Jena with the seventy-year-old Goethe[53] – and that of nicotine from tobacco in 1828. In 1832 a separate alkaloid, codeine, was extracted from opium, and in 1842 theobromine was isolated from chocolate. These discoveries transformed chemistry, particularly in Germany, into an industrial science that identified new plant-derived compounds and supplied them at scale to a vast pharmaceutical market.

Self-experiment with drugs also informed the nascent study of the mind, nourishing the roots of what would become psychology and neuroscience. In 1819 the Czech physiologist Johann Purkinje was embarking on his famous studies of entoptic or 'Purkinje images', in which he extended the self-experiments of Newton and Ritter by mapping the phenomena of subjective vision. He began by recalling a game from his childhood, in which he faced the sun with eyes closed and passed his spread fingers in front of them to generate strobing colours. He proceeded to enumerate and classify the after-images, phantoms, pinwheels, haloes and blind spots that could be generated by similar methods, tracing each one back to their physical causes in reflections from the cornea, the lens and the blood vessels in the eye.

Over the course of his long career, during which he became professor of the world's first university department of physiology at Breslau, Purkinje coined the word 'protoplasm', described animal cells and their nuclei for the first time, identified gastric glands and sweat ducts, and witnessed the fertilisation of the ovum by sperm. His studies of the grey area between physically observable and subjective effects made him fascinated by mind-altering drugs and plants. He noticed that belladonna produced physical changes in the eye and administered it

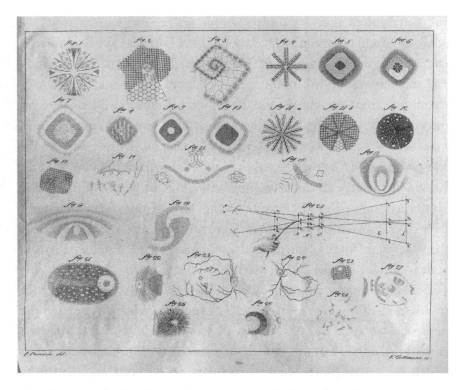

4. Johann Purkinje's chart of subjective visual effects, from his 1823 doctoral dissertation.

to himself, using a special viewing tube to magnify the visual distortions and chromatic patterns that formed as the drug relaxed the iris and the lens scattered the light. He was the first to record the effects of digitalis on vision, drawing the patterns that danced before his eyes after drinking a decoction of the toxic leaves. He experienced a more profound mental derangement when he ate three nutmegs in Berlin and took a stroll to the Royal Theatre:

> The distance was long, but this time I thought it had *no end*. My movements appeared entirely adequate, but were lost momentarily in dream pictures, from which I had to extricate myself with consid-

erable force in order to keep on walking. My feet did their duty, and, since I had to stick to a straight road, there was no danger of going astray. I went forward in this dream, for, if I attempted to orient myself, I could not even recognise the cross streets. Time seemed long. I got to the opposite side of the place where I was going. During this time dreams and physical activity battled one another.[54]

Purkinje followed this with a further experiment, in which he dissolved 8 grams of nutmeg in brandy and drank the solution. This time the effect was quite different: he was dizzy, his muscles twitched spontaneously, and a long walk was out of the question. One drug appeared to change the action of the other: the oils in nutmeg acted differently on the motor functions when they were dissolved in alcohol. Purkinje discovered that drug interactions were complex and unpredictable, and believed they held the key to many physiological processes that were otherwise inscrutable. Unlike Davy and Ritter, however, his aim was to eliminate the personal reflections in his self-experiments and to transform them wherever possible into external and measurable data. Ideally, he wrote, experiments of this kind would be carried out on large numbers of people, to generate an average response that minimised individual subjectivity and variation.

* * *

Purkinje's injunction that the study of drugs should become more objective and data-driven would be widely taken up in the generation to come. But the opposite tendency was also on the rise, as the literature of self-experiment extended beyond the boundaries of science. Just as Humphry Davy had deployed the Romantic poets' 'language of feeling' to scientific ends, embracing the aesthetic and sublime dimensions of the drug experience, Romantic literary voices now turned the language of science and medicine to their own purposes. The most influential, Thomas De Quincey, was the protégé of Samuel Taylor Coleridge, who had forged a passionate friendship with Humphry Davy after participating in his nitrous oxide experiments in 1799.[55] De Quincey, a teenage acolyte of Coleridge and Wordsworth, had become Coleridge's secretary at a time when he was heavily dependent on opium, and he was a frequent witness to the private indulgences and prostrations that his mentor concealed from public view. In 1820 De Quincey, opium-haunted himself and with editors and creditors chasing unwritten articles and unpaid bills, offered his personal perspective on the subject to the new and generously funded *London Magazine*. It was a last throw of the dice for his faltering career, but the ensuing book, *Confessions of an English Opium-Eater*, was a sensational success that kept him afloat for the next fifty years.

When a volume of his self-experimental reports was proposed, Johann Ritter had insisted: 'It should not merely be a collection of my writings; it will be a kind of literary autobiography, of interest perhaps to anyone who wants or has to educate himself to become a physicist and experimenter.'[56] De Quincey took a similarly autobiographical approach to describing opium: before you can understand the drug, he insisted, you must understand its subject, the opium eater. His *Confessions* presented itself both as a minutely observed personal history and as a treatise from a uniquely qualified medical expert. 'Upon all that has hitherto been written on the subject of opium,' he announced in its opening pages, 'I have but one emphatic criticism to pronounce – Lies! lies! lies!'[57] His narrative encompassed his entire life story, but it was also:

the doctrine of the true church on the subject of opium, of which church I acknowledge myself to be the only member – the alpha and omega: but then it is to be recollected, that I speak from the ground of a large and profound personal experience: whereas most of the unscientific authors who have at all treated of opium, and even of those who have written expressly on the materia medica, make it evident, from the horror they express of it, that their experimental knowledge of its action is none at all.[58]

De Quincey's descriptions of what would by Freud's time be diagnosed as 'addiction' – the pleasures and pains of craving, tolerance and withdrawal – were far in advance of most doctors of his day. Yet his claim to scientific authority was also ironic, a facetious riposte to medical pretensions of omniscience. Opium was a familiar drug, not just to doctors but to the public at large: it was by far the most effective painkiller available, and was sold freely in markets, groceries and pharmacies as a sovereign remedy for everything from headaches to rheumatism, diarrhoea and menstrual cramps. In reconceiving it as the source of exquisite pleasures and pains, De Quincey was in part playing a Byronic game, advertising himself as the champion of a new vice, baiting the moralists and delighting the jaded wits. His claim to the authority of science was made in the same spirit: if doctors dismissed him as unqualified, he answered them with science's own motto, 'Nullius in Verba'. His originality, and the reason for his tremendous influence over the following decades, was in the way he used the drug as a device for exploring the hidden recesses of his mind:

A theatre seemed suddenly opened up and lighted within my brain, which presented nightly spectacles of more than earthly splendour . . . I seemed every night to descend, not metaphorically, but literally to descend, into chasms and sunless abysses, depths below depths, from which it seemed hopeless that I could ever reascend.[59]

De Quincey dredged up from these hidden depths not merely details that had been beyond his conscious recall, but entire worlds whose existence the rational intelligence never suspected. The philosophers of the eighteenth century had championed the powers of reason over the irrational; but De Quincey had read deeply in the recent countertradition, in Germany in particular, that proposed the existence of a hidden or 'unconscious' mind of awesome depth and potency. Some philosophers identified this buried inner world with nature, the boundless ocean in which an insignificant humanity swam. Others glimpsed it in the creative process, in which inspiration and art seemed to well up fully formed without effort or volition. Friedrich Schiller, whom De Quincey read avidly and who would exert a powerful influence on both Freud and Carl Jung, believed that authentic poetry should be unmediated by the conscious, rational, tinkering mind; Johann von Goethe claimed to have written his novel *The Sorrows of Young Werther* (1774) while in a daze or trance, 'practically unconscious'.[60]

In his *Confessions*, De Quincey describes how he used opium to enter this mysterious domain at will and to assemble his life story anew from the fragments he retrieved from its depths. During his drug reveries, his mind opened onto a maze of secret chambers, trapdoors and passages, vast and shadowy and sublime as the vaults and dungeons of Piranesi's etchings that Coleridge had once memorably described to him. In his opiated mental theatre, mythic dreamscapes were woven together with forgotten scenes from childhood, claustrophobic dreamscapes and cameo scenes of waking life glimpsed from the roof of a stagecoach or the crowded mêlée of Covent Garden on a Saturday night. It was an incomparably vivid portrayal of the life of the mind and the power of drugs, in which medicine and science were consigned to the margins.

In his reworked and expanded edition of the *Confessions* of 1856, De Quincey was no longer the only member of the true church of opium: he had promoted himself to its Pope. By that time his artfully constructed alter ego, the Opium-Eater, had taken on a cosmopolitan life of its own. In Russia, Nikolai Gogol had adapted his London dream-wanderings to his own St Petersburg in *Nevsky Prospect* (1835); in America, he had been extolled by Ralph Waldo Emerson and imitated by Edgar Allan Poe. In France, Alfred de Musset's freely elaborated translation had inspired Théophile Gautier and Honoré de Balzac to create their own versions of the new archetype, and Hector Berlioz to transpose its nightmare fugues to music in his *Symphonie*

Fantastique (1830). By the time of De Quincey's death in 1859, Charles Baudelaire was at work on his translation of the *Confessions* that, after its appearance in *Les Paradis artificiels* (1860), would become a ubiquitous reference point for the fevered drug culture of the *fin-de-siècle*.

* * *

Within science, meanwhile, self-experimentation with drugs was being challenged and marginalised by a new generation of thinkers. In France, the positivist philosophy of Auguste Comte inspired many who wished to reshape the study of the mind by stripping it of metaphysical speculation. Mental events, Comte argued, were not facts, and a 'science of the mind' was an oxymoron. 'Direct contemplation of the mind by itself is pure illusion,' he wrote in 1830: its expert practitioners disagreed fundamentally on its most basic precepts, and there were no possible checks or references that could resolve their disputes.[61] The mind observing itself simply devolved into an infinite regress, speculation piled on speculation.

In Germany, a new generation of physiologists had begun to investigate the mechanics of sensation and perception with precision instruments. These realised Johann Purkinje's aspiration that the study of subjective experience could be advanced by systematic trials and measurements, reducing the vital forces, fluxes and life-sparks hypothesised by *Naturphilosophie* to the material laws of physics and biology. The physiologist Hermann von Helmholtz, combining his expertise in mechanics, biology and chemistry, demonstrated that energy could be accurately measured and transposed across different forms – chemical, electrical, magnetic – and began to map the microscopic nerve structures that conducted it through animal tissue. In 1851 he invented the ophthalmoscope to inspect the tiny structures that generated Purkinje's visual hallucinations. Helmholtz and his followers, who included Freud's professor and mentor Ernst Brücke, explicitly rejected metaphysical speculation about immaterial life-forces and held firm to the principle that 'no other forces than the common physical-chemical ones are active within the organism'.[62]

In 1862 another pioneering physiologist, Wilhelm Wundt, set up a laboratory to test human reaction times and identify the split-second processes that marked the transformation of raw sensation into conscious perception. In 1879, working from his experimental laboratory at the University of Leipzig, Wundt became the first person to call himself a 'psychologist'. In his mechanistic paradigm, the study of the mind shifted away from self-reports by a single subject-observer and towards a battery of brass precision instruments able to measure phenomena beneath the threshold of conscious awareness.

As science professionalised and became more technically complex, it created new expectations of the scientist.[63] Modern practitioners required a set of skills and concepts quite different from the intuitive leaps and poetic observations of the Romantic generation that had preceded them. By mid-century, these were gathered under the umbrella term 'objectivity': a new concept, related to but more precise than long-established scientific terms such as 'truth' and 'certainty'. It derived from Kant's philosophy, in which the noumenal world was subjective and thus beyond the scope of science, and science was the study of 'objective', measurable phenomena. By mid-century, 'objective' had made its way into common speech. When Thomas De Quincey used it in his revised 1856 edition of *Confessions of an English Opium-Eater*, he glossed it with the comment:

> This word, so nearly unintelligible in 1821, so intensely scholastic, and consequently, when surrounded by familiar and vernacular words, so apparently pedantic, yet, on the other hand, so indispensable to accurate thinking, and to wide thinking, has since 1821 become too common to need any apology.[64]

Objective science was quantified and data-driven; it demanded patience, precise observation and accurate measurement. It placed a premium on the skills, judgement and sensibilities of a scientifically trained observer who knew which facts to select and which anomalies were significant, and recorded them in well-ordered notebooks. In the study of drugs, this privileged the careful measurement of externalities – dosage, times of onset, recording of symptoms – over what was now described as 'introspection', which was liable to shape mental phenomena into narratives pleasing to the self-observer. Creativity of this kind was the wellspring of art, but it had no place in positivist or materialist science. The first generation of psychological researchers, anxious to distinguish themselves from literary pretenders to experiment such as De Quincey and to produce results that fitted the new demands and standards of objectivity, approached the study of drugs with self-conscious sobriety.

Who counted as a trained observer, and what did the training consist of? A growing number of practitioners were educated at universities and medical schools, but an academic degree was still far from a necessary qualification. The distinction was as much cultural as technical, rooted in the informal 'Cartesianism of the genteel'. Its criteria were rarely spelled out explicitly, but Baron Ernst von Bibra – himself a wealthy amateur rather than the product of a university – provided a telling example in 1855, while relating his first experiment with hashish. He had swallowed 14 grains (roughly a gram) of Egyptian '*charas*', noting its 'bitter, resinous, but not disgusting taste', and some two hours later found himself under the influence of an effect much stronger than he had anticipated. Staring at a panelled wall, he saw the windows of his study transposed into it, and noticed that he could ' "move" that part of my room to a distance and thus extend my study into an infinitely large apartment'. Equally, however, 'I could at my will make all these delusions disappear and also suppress my hilarity'. He concluded: 'If excited by an overdose of hashish or opium, a savage Moslem becomes an assassin, a Malay is seized by an amok frenzy, whereas in the same situation, a scholarly, educated European medical doctor carries out observations on himself.'[65]

From its beginnings in the Royal Society, experimental science had drawn heavily on knowledge accrued in other parts of the world, but it had always been necessary to recast it in terms that transformed it from colourful traveller's tales to testable scientific data. When, for example, Robert Knox passed his sample of '*bangue*' to Robert Hooke in 1672, he told his friend that 'they call it in Portugeze *banga*', implying that it was well known to pharmacists and apothecaries elsewhere in Europe.[66] There were many representatives of these trades in Lisbon who could have described the effects of the drug, not to mention any number of Sri Lankans from where Knox had sourced the drug, but Hooke presented it to the Royal Society as a novel discovery. Before drug experiences were admitted to the status of scientific knowledge, they needed to be reported in English by a trained observer.

The criteria for objective reporting were gendered as thoroughly as they were racialised. Formal higher education was almost exclusively for men, and medical science remained an overwhelmingly male domain until well into the twentieth century. Within its sphere, women were rarely used as experimental subjects, particularly with drugs that might result in disinhibition and loss of decorum. This was a risk of which women were acutely conscious, and it had been noted in the first experiments with nitrous oxide. Thomas Beddoes was an outspoken proponent of women's rights who, at the outset of Humphry Davy's experiments in 1799, had just given Bristol's first course of public science lectures for women. He and Davy initially aimed for a balance of the sexes among their volunteers, but the experience of the first female subject, who turned into a 'temporary maniac', rushing out of the laboratory and sprinting down the street, 'so intimidated the ladies, that not one, after this time, could be prevailed to look upon, or hear of nitrous oxide, without horror!'[67] Public displays of disinhibition were easy enough for male subjects to laugh off as a flamboyant performance or frame with hindsight as the heroic pursuit of knowledge; but no woman could, as Humphry Davy did, strip to the waist in an airtight chamber filled with nitrous oxide and launch herself into an uncharted dimension of pure thought.

Not all self-experiments were public, but all had a public dimension, since they had to be capable of demonstration and replication. As the historian Naomi Oreskes has observed, the opposing poles of nineteenth-century experimental practice – heroism and objectivity – trapped female researchers in a double bind. Objectivity was also perceived as a masculine quality, demanding the ability to view data dispassionately rather than colouring it with feeling.[68] As a result, female drug experiences were largely confined to medical case histories, patient testimonies and, on rare occasions, being co-opted into the experiments of their male relatives.[69] Self-experimenting women were more often to be found outside the world of science, for example in literary or occult circles, though even here their practices were often undocumented, since public confession left their subjects exposed to potential scandal. The first professional female scientists became involved in drug experiments only in the twentieth century, and even then they were likely to be targeted and 'exposed' by the press.

*   *   *

The scientific world that the young Freud inhabited at Vienna University was dominated by the disciples of Helmholtz and Wundt, and the startling progress they were making in mapping the mechanics of sensation and perception. Yet the effects of drugs on the mind, and the biological and nervous processes that lay behind them, remained a profound mystery. Wundt tended to regard them, along with dreams and madness, as phenomena unsuitable for objective study. Outside the laboratory, however, and particularly for physicians who saw medicine as an art as much as a science, the subjective responses elicited by 'How do you feel?' were indispensable. The popular and wide-ranging survey of pharmacy by James Johnston, *The Chemistry of Common Life* – first published in 1855 and regularly reprinted for decades – observed 'how very defective our knowledge is, both of the chemical nature and of the physiological action of the narcotics in which we indulge', and urged that this defect should be addressed by the only currently effective method: 'direct experiment, on man'.[70]

Many specialists in the study of mind-altering drugs rejected the strictures of positivism and the sober demands of the new scientific self. They saw their field of study as a particular case in which the most significant data were irreducibly private and subjective. This position was argued most eloquently and influentially by the French physician Jacques-Joseph Moreau, whose 1845 treatise *Hachich et l'aliénation mentale* (*Hashish and Mental Illness*) remained the most authoritative treatment of the mental effects of cannabis for the rest of the century. Moreau announced on the opening page, 'I had become acquainted with the effects of hashish through my own experience, and not merely from the reports of others', and went on:

> Indeed, there is essentially only one valid approach to the study: observation, in such cases, when not focused on the observer himself, touches only on appearances and can lead to grossly fallacious conclusions. At the outset I must make this point, the verity of which is unquestionable: Personal experience is the criterion of truth here. I challenge the right of anyone to discuss the effects of hashish if he is not speaking for himself and if he has not been in a position to evaluate them in light of sufficient repeated use.[71]

Moreau's experiments with hashish involved large oral doses that plunged him into hours of hallucination, an experience that took him far beyond the limits of objectivity and the disembodied stance of the trained observer. He challenged the presumption that self-experiment was compromised by its dual nature; on the contrary, he insisted, drugs that affect the mind demanded an investigator who occupied the roles of both observer and subject. An observer might notice that a subject who had consumed a large dose of hashish was supine, unable to move or speak, and assume that they were sedated or semi-conscious; anyone who had taken the drug themselves, however, would recognise that the reason they were unable to converse was that their mind was racing too fast to form and express coherent thoughts. As a practising alienist – he was resident physician at the Bicêtre Hospital in Paris – Moreau was acutely aware of the limits of external examination with mentally disturbed patients. 'We see only the surface of things,' he wrote; 'can we be certain we are in a condition to understand these sick people when they tell us of their observations?' The experience of hashish, he argued, offers a privileged glimpse of the world as it appears to those whose mental processes are similarly disordered: 'To understand an ordinary depression, it is necessary to have experienced one; to comprehend the ravings of a madman, it is necessary to have raved oneself, but without having lost the awareness of one's madness.'[72]

Moreau came to believe that the most effective use of hashish in mental therapy was to give it not to the patients, but to their doctors. It could provide them with the *sens intime*[73] – as we might say today, the lived experience – of mental phenomena such as obsession, depersonalisation or delusion, allowing them to understand the conditions they were treating by experiencing them in themselves.

* * *

As he wrote 'Über Coca', Freud sat at the confluence of these two traditions. He was attempting to weave objectivity together with introspection, and combine the strengths of both approaches. His professor at Vienna, Ernst Brücke, sat squarely in the objective tradition, a full-throated member of the group of committed materialists who referred to themselves as the Helmholtz School of Medicine. But it was the introspective school of self-experiment that had generated the most penetrating descriptions of cocaine's action, reaching far beyond the limited language of neuroanatomy. By writing in the first person, Freud had enlisted himself the heroic tradition of discovery that included Davy, Moreau and Mantegazza; but his habitual caution meant that he outsourced the heroism to others. 'The effect of large doses of coca', he wrote, 'was investigated by Mantegazza in experiments on himself';[74] Freud himself never exceeded his initial dose of 50mg, and had found that 'a first dose or even repeated doses of coca . . . produce no compulsive desire to use the stimulant further'.[75] Mantegazza was, he acknowledged, an 'enthusiastic eulogist' for the drug, whose report had 'aroused much interest but little confidence' among medical professionals; Freud judged, however, that 'I have come across so many correct observations' in his report that 'I am inclined to accept his allegations'.[76]

At this point he began taking small doses of cocaine regularly outside of his research and found it useful, enjoyable and fascinating in equal measure. He would confess to Martha that 'my tiredness is a sort of minor illness; neurasthenia, it is called';[77] with cocaine, however, he soon noticed 'the disappearance of elements in one's general state of well-being that cause depression'[78] and by June, he wrote to Martha, he was 'strong as a lion, happy and cheerful'.[79] Cocaine, he also discovered, was a performance enhancer. 'One feels' – the impersonal pronoun situating his claim somewhere in between personal experience and objective data – 'more vigorous and more capable of work.' Cocaine 'steels one to intellectual effort', in which 'long-lasting, intensive mental and physical work can be accomplished without fatigue'.[80] Significantly, he found that this increase in nervous energy, unlike that produced by alcohol or caffeine, was 'not followed by any feeling of lassitude or other state of depression'. If Beard was right that the many manifestations of neurasthenia were symptoms of a single underlying cause, cocaine might be the sovereign remedy for it that so many had been seeking.

But cocaine had yet another quality: as well as an anti-depressant and an energy booster, it was a powerful euphoriant. Freud quoted Mantegazza's claim that it produced a 'state of greatly increased happiness':[81] during the peak of his coca intoxication he had scribbled, 'God is unjust because he made man incapable of sustaining the effects of coca all life long. I would rather have a lifespan of ten years with coca than one of 1000000000 centuries without!'[82] In Mantegazza's view, this was the drug's primary effect, and he had gone on to construct his expansive vision of a utopian future society structured around ecstatic inebriation. Similar observations recurred in the literature on other intoxicating drugs, from opium to nitrous oxide to hashish; Moreau, for example, had identified happiness as the primary manifestation of hashish intoxication:

> At a certain moment in the intoxication, when an unbelievable effervescence takes possession of all the mental faculties, a psychic phenomenon is evident, the most curious of all perhaps, and one that I despair of conveying appropriately. It is a feeling of physical and mental well-being, of inner contentment, indefinable joy, impossible to analyse, to understand or to explain . . . Following this happiness, which is so agitated that all your being shakes convulsively, comes a gentle feeling of physical and mental lassitude, a sort of apathy, insouciance, a complete calm, to which your mind abandons itself with delight.[83]

If there was such a thing as a seat of happiness in the brain, and there were drugs able to stimulate it at will, this would be a discovery of enormous consequence. It would also be fraught with danger: what would stop drug users simply pursuing this pleasure at all costs, at the expense of their fellow citizens and their sanity alike? As an alienist, Moreau was well aware that happiness and laughter could be seen as pathological, maniacal or hysterical; hence his insistence that the happiness of hashish was, despite its extravagance and irrationality, the expression of a rational and healthy mind. Humphry Davy and his circle had similarly observed that subjects would often laugh uproariously under the influence of nitrous oxide, yet be unable moments later to explain what had prompted their hilarity: a response that would normally be taken as a symptom of mental disturbance, but it arose repeatedly in perfectly healthy volunteers. How could such profound happiness arise from no other cause beyond the gas itself?

For Mantegazza, the ecstasy of cocaine was an altered state of mind that had the potential to reorient its subjects towards a revolutionary society constructed around pleasure. Freud, by contrast, was careful to present its mood elevation as essentially moderate, stressing that 'the exhilaration and lasting euphoria' that manifests after 50–100mg of cocaine 'does not differ in any way from the normal euphoria of a healthy person'.[84] The emphasis of 'normal' and 'healthy' served two purposes: first, to stress that cocaine did not alter the personality or threaten the sober self and, second, to counter the unhealthy associations that the term 'euphoria' had begun to acquire. Originally, euphoria meant simply feeling well: the experience of health, as described by the patient. Freud understood it in these terms as a positive symptom, a sign that the needs of body and mind were satisfied, like the purring of a well-tuned engine. Other medical voices, however, had begun to conceive it as a danger sign, analogous to the paradoxical feeling of relaxation and contentment sometimes described by those rescued from the last stages of drowning. It was particularly associated with the later stages of consumption, where patients were often described as exhibiting a 'euphoria or morbid hopefulness' as their condition worsened.[85] The French physician Charles Féré, in a survey of the condition in 1892, claimed that 'euphoric crises' could be found in epileptics and hysterics, and regarded them as a dangerous dissociation from the body that led patients to mistake their illness for health.[86] Euphoria, in this recent medical usage, was pleasure with an asterisk. When produced by an unnatural stimulus such as a drug, it might be the delusion of a subject who, beneath the sensation of pleasure, was in reality being poisoned.

It might have been politic for Freud to play down cocaine's euphoric effects, but his interest in them extended further. He wondered, as Moreau had with hashish, whether the euphoria it induced was in some sense its primary effect, rather than merely a side-effect of its stimulant properties. Could it be that the drug induced a mental shift that allowed the nervous system to access untapped reserves of energy?

Just as the visions of opium or hashish were not contained in the drugs themselves but in the otherwise inaccessible dimensions of mind to which they allowed access, the increase of physical and nervous energy on cocaine might be the product of a mind optimised to a higher pitch of functioning.

If this hypothesis was true, it should be measurable, and Freud turned next to the brass-instrument experimentation favoured by his professor, Ernst Brücke. Euphoria was a subjective sensation, but mental or physical energy generated by it should have objectively measurable correlates. Cocaine, as Freud had already learned from trying it on a handful of friends and colleagues, had different effects on different subjects; it should therefore be possible to show whether the increase in physical energy correlated simply with the dose administered, or whether those who experienced greater euphoria also evinced a more powerful stimulant response. To test this, Freud had the use of two experimental devices: a dynamometer, which measured the pressure exerted on it and locked its needle at the point of maximum pressure, and a neuroamoebimeter, a vibrating metal strip that registered reaction times. It was the first and, as it turned out, the only time that Freud ever experimented on living human subjects: predominantly on himself, as his response to cocaine seemed more predictable and more positive than those of the volunteers he enlisted.

In January 1885 he published his results in a short paper, 'Contribution to the Knowledge of the Effect of Cocaine'. From his tabulated columns of dates, doses, mechanical pressures and reaction times he was able to demonstrate that the subjective sense of increased strength and energy induced by cocaine was objectively real: when on the drug, he and his volunteers exerted more pressure on the dynamometer, maintained it for longer and had faster reaction times. He was also able to demonstrate that this increase in energy began as soon as the rush of euphoria was felt, which was before most of the drug had been absorbed into the bloodstream. From this he concluded that cocaine's stimulant action was not produced directly by the nervous system but was 'indirect, effected by an improvement of the general state of well-being'.[87]

Freud's hybrid methods, like his hybrid literary style, had produced a striking outcome: he had used physical measurement to track a subjective alteration in consciousness, and produced objective data to suggest that the mind was the source of a physiological effect. Around this time he switched from dissolving his cocaine in water to sniffing the powder. This was a far more cost-effective route: the drug is partially broken down and rendered inactive in the stomach, but the nasal membranes relay it directly into the bloodstream. Sniffing also makes the onset of its effects more rapid and pronounced.

By now medical and pharmaceutical interest in cocaine was on the rise, and other physicians were beginning to hail it as a miracle drug. Parke, Davis were expanding their supply and range of preparations; John Pemberton, a patent medicine entrepreneur in Georgia, announced a new cocaine-based beverage, Coca-Cola, touting it as 'a great invigorator of the brain'.[88] 'Über Coca' was reprinted, and Freud, whom interested doctors and pharmacists now considered the leading medical authority on the drug, was working on an expanded version. His ingenious experiments had squared the circle of objectivity and introspection; his innovative writing style captured the subjective sensations of cocaine while his laboratory work underpinned them with quantitative data. He had in his sights a cure for the nervous disease of the age, a performance enhancer to correct the growing disparity between the powers of the human mind and the demands of modern life, and the world was racing to catch up with his discoveries. What could possibly go wrong?

CHAPTER TWO

PROSTHETIC GODS

Among the colleagues Freud had enlisted in his dynamometer experiments was a young intern in Vienna Hospital's ophthalmology department, Karl Köller. As they swallowed their doses of cocaine in water, both noticed the instantaneous numbing sensation around the lips and throat that Freud had described after his first experiment. Köller, however, was particularly attentive to it. His consuming professional interest was in optical physiology – to the extent that Freud privately found him rather a bore – and he grasped immediately that this might make cocaine invaluable for eye surgery. Even the most powerful analgesics were unable to prevent reflex twitching and blinking, making common procedures such as removing cataracts almost unendurably painful. Köller's hunch was easily validated: he dripped cocaine solution into a frog's eye and established that he could touch its bulging cornea without making it blink. He promptly arranged a demonstration where, in the presence of witnesses including a senior ophthalmologist, he repeated the procedure with a dog.

Köller's paper on his discovery was read before the Vienna Medical Society on 17 October 1884. 'Cocaine', he announced, 'has been prominently brought to the notice of Viennese physicians by the thorough compilation and interesting therapeutic paper of my hospital colleague Dr Sigmund Freud.'[1] Freud was initially thrilled that 'a colleague has found a striking application for coca in ophthalmology': his new discovery was on the march. But it gradually became clear that this was to be cocaine's only uncontroversial medical application, and that Freud had been relegated to a minor supporting role. He had been the first to propose it in print: 'Über Coca' concluded with a series of therapeutic suggestions for the drug, the last paragraph of which had noted that its 'marked anaesthetising effect when brought into contact with the skin' might lead to 'a good many further applications'.[2] As Köller's career took off, Freud began to wonder why he himself had not taken the step, obvious with hindsight, of applying his suggestion to surgery. On occasion he cursed himself for his laziness; later, unchivalrously, he blamed Martha's distracting influence. Forty years later, he concluded that the root cause was that cocaine had ultimately been a distraction from his specialism in neurology and he had never properly committed his attention to it. To the extent that his investigation of cocaine had been an opportunistic pursuit of career advancement, it was ironically fitting that the prize should go to another. Köller rode his success to a lucrative ophthalmology practice in New York; in later life Freud drolly referred to him as 'Coca Köller'.

The episode carried a curiously precise echo of Humphry Davy's nitrous oxide researches. In his *Researches Chemical and Philosophical* (1800), Davy had made a very similar suggestion, that the gas was 'capable of destroying physical pain' and consequently could 'probably be used with great advantage during surgical operations'.[3] But the suggestion went no further, and by the time nitrous oxide anaesthesia emerged fifty years later it had been all but forgotten. Both Davy and Freud were absorbed by the novel states of consciousness they were exploring, and the profound questions they raised about the nature of the mind and its relations to the body – specifically, in both cases, the nature of the connection between chemical stimulus and pleasure. Both were interested in measuring and recording physiological data, but their primary focus was introspective, mapping previously unexplored landscapes of thought and sensation.

Freud may have been slow to appreciate the importance of Köller's discovery, but it acted as an immediate spur to the nascent cocaine industry. In the US, Parke, Davis ramped up production and by 1887 they were one of several suppliers offering the pure drug at wholesale prices that had dropped from a dollar a grain to as little as 2 cents.[4] Six months after the publication of Köller's paper, the price of Merck cocaine, already high at 6 marks per gram, had quadrupled, and Merck followed Parke, Davis in importing crude cocaine extracts from the Andes instead of shipping unprocessed coca leaves in bulk to Europe. Under pressure from the US competition, Merck justified their price hike by advertising that their cocaine was of the highest quality, a claim they supported with Freud's use of their product in 'Über Coca'.[5]

It was another of Freud's therapeutic suggestions, however, that tipped his cocaine researches from promise to disaster. As part of his case that it was a remedy for nervous weakness and depression, he relayed the claim with which Parke, Davis had originally launched the drug in the USA that it was 'an antidote to the opium habit', removing cravings and the pains of withdrawal.[6] The evidence for this, as Freud had noted, was largely drawn from the *Therapeutic Gazette*, Parke, Davis's in-house pharmaceutical journal, which now regularly published case histories and testimonials of successful opium cures. These had been a significant spur to Freud's own researches, not least because he had an experimental patient in mind.

In the letter to Martha in which he first announced his interest in cocaine, he wrote that he intended to try it in cases of heart disease and nervous exhaustion, and 'particularly in the awful condition following withdrawal of morphine (as in the case of Dr. Fleischl)'.[7] Ernst von Fleischl-Marxow was a brilliant older colleague of Freud, a junior professor under Ernst Brücke at Vienna Medical School, where he had made pioneering studies of electrical activity in the nerves and brain. Freud idolised him; as he wrote to Martha, 'I admire and love him with an intellectual passion, if you will allow such a phrase.'[8] For years Fleischl had lived with a terrible injury. He had infected his right thumb with a scalpel during an autopsy, and the subsequent amputation left him with nerve damage and constant, often excruciating pain. He bore his condition stoically and as gracefully as possible, but by

1884 he was relying on morphine to manage it, injected in ever larger doses. His escalating drug habit, combined with the pain it only partially suppressed, was destroying him.

Fleischl was one of the first to hear of Freud's discovery of cocaine; he was enthusiastic, and asked Freud to try the cocaine cure on him. The effects were instantaneous: the pain disappeared, the cocaine lifted his spirits and he was quickly able to reduce and then dispense entirely with his morphine. Fleischl wrote a short note on the success of his case, in which he theorised that opiates and cocaine were somehow antithetical to one another. Freud, writing 'Über Coca' at the same moment, included the case of an anonymous patient who had successfully substituted morphine for cocaine and 'after ten days was able to dispense with the coca treatment altogether'.[9]

But the miracle was short-lived. Within a week, Fleischl was using cocaine in quantities Freud had never imagined was possible. Over the following three months, he returned to morphine while also using at least a gram of cocaine a day, taking it by subcutaneous injection. He was spending a hundred times more on it than Freud did even during his periods of most regular use. He developed insomnia, paranoia, and a nerve-shredding delirium in which he felt snakes crawling all over his skin. Freud spent harrowing nights with him in which Fleischl talked incessantly and crazily: 'every note of the profoundest despair was sounded', leaving Freud wondering 'if I shall ever in my life experience anything so agitating'. Pain, exhaustion, morphia and cocaine: 'all that makes an *ensemble* that cannot be described'.[10]

This was a dimension of cocaine that Freud's cautious self-experiments had entirely failed to uncover. Confronted every morning with the embroidered injunction 'If in doubt, abstain' on his wall, he had never felt the urge for more than 'an effective dose' of 50mg, nor to follow one dose with another. Rather, he had noted 'a slight revulsion' at the prospect of taking a second dose before the first had worn off.[11] He had relied for his descriptions of high doses on Mantegazza, who 'experienced the most splendid and colourful hallucinations, the tenor of which was frightening for a short time, but invariably cheerful thereafter';[12] Mantegazza, however, had been chewing coca leaves, which set a practical ceiling on the quantity of cocaine he absorbed. Depending on the potency of the leaf, his most extravagant doses probably amounted to around a gram spread over a day and an evening; and his method of gradual ingestion, combined with the other alkaloids and minerals in the leaf, muted the nervous effects that Fleischl's method of injection heightened. Freud's experiments on animals had shown that high doses produced undesirable physical effects – rise in pulse and blood pressure, gastric upset – and he assumed that, as with caffeine, higher doses would prove self-limiting.

Freud was, however, heavily invested in his belief that cocaine was a miracle drug, a belief now shared by many of his medical colleagues. In March 1886 *Chambers's Journal*, a popular British review of the arts and sciences, described it as a 'discovery which has surpassed the ordinary standard of greatness sufficiently to enable it to figure as one of the wonders of the age'. In 1884, with Freud's support, it 'flashed like a meteor before the eyes of the medical world, but, unlike a meteor, its impressions have proved to be enduring'.[13] It was being marketed in various forms – a sniff or a syrup, a lozenge or a herbal cigarette – for a huge range of conditions, and its public reputation and image reflected the enthusiastic claims of its promoters. Its customers found symptomatic relief for bronchial and sinus conditions, and an elevated mood into the bargain. In a modern world where fatigue and depression were endemic, it promised miraculous relief from previously intractable and chronic suffering.

It was equally popular as a mental stimulant, and doctors were among the categories of brain-worker to whom it particularly appealed. In October 1888, an article in the *Edinburgh Review* connected cocaine to the heroic role required of medicine in the modern world. Doctors, it argued, were busier than ever, with complex and demanding work that afforded limitless stimulation 'to the most enlightened and far-reaching mind'. The modern practitioner 'has scope for muscular exercise; he has always to be acquiring new information, which keeps the mental organism employed'. In all these respects, cocaine was 'invaluable as an internal remedy', boosting not only stamina and mood but intellectual capacity.[14] The vivid portrait Freud presented of the cocaine-enhanced mind and body, firing vigorously on all cylinders to meet the demands of the modern world, was a perfect fit with the self-image of the modern medical profession.

* * *

Fleischl's case was a medical first, though not in any sense that Freud wished for. It was soon joined by others. In July 1885 the neurological journal *Centralblatt für Nervenheilkunde* carried a commentary by its editor, the physician and asylum superintendent Friedrich Albert Erlenmeyer, describing the novel condition of cocaine addiction. Erlenmeyer was the author of one of the earliest books on morphine addiction and its treatment, and he paid close attention to new and emerging forms of drug craving. The following year he wrote another article for the *Centralblatt* pronouncing cocaine the 'third scourge of humanity', after alcohol and opium. The third edition of his book included a short passage crediting Freud as a pioneer of cocaine thera-peutics and adding that 'he recommends unreservedly the employment of cocaine in the treatment of alcoholism'.[15] It was a difficult accusation to contest, especially once Ernst von Fleischl-Marxow had become the first patient to be diagnosed with a dual addiction to morphine and cocaine, acquired while under Freud's supervision.

Despite opium's great antiquity, addiction was a modern diagnosis. It had been noted by classical Greek and Roman physicians and phar-macists that those who used opium regularly were obliged to increase their dose, and it was fifty years since Thomas De Quincey had spelled out the agonies of withdrawal in unsparing detail. Yet this factor remained of less concern to doctors than the more acute danger of opium, its narrow dosage window: only two or three times the effective dose could be enough to induce a potentially fatal respiratory depression. But opium – and, after around 1850, its more potent synthetic extract, morphine – were too valuable for pain relief for doctors to avoid them on these grounds. There were plenty of other drugs that needed to be taken every day, and chronic use was a lesser concern in an era when opium was widely available and relatively inexpensive.

In many respects Fleischl was typical of the first cohort to be diagnosed as 'narcomaniacs', later to be known as 'drug addicts'. The condition was originally described by physicians who specialised in nervous diseases and offered residential care to private patients, and was yet to become familiar outside this milieu. It was first formulated during the 1870s as 'morbid craving' by doctors such as Eduard Levinstein, director of one such institution in the Schöneberg district of Berlin. In 1877 Levinstein published the first book on the morbid craving for morphine ('Morphiumsucht'), describing its symptoms with reference to the case notes of his patients, supplemented with animal experiments on dose and toxicity. He saw the craving not as a mental disease, but a novel addition to 'the category of other human passions, such as smoking, gambling, greediness for profit, sexual excess etc.'.[16] He was equally clear that it was not an intellectual deficit: in fact, many of his patients were 'authorities in military matters, artists, physicians, surgeons, bearing names of the highest reputation'.[17] If anything it seemed to select its victims chiefly from the upper echelons of society – as, of course, private clinics such as Levinstein's did.

Addiction, in his view, was ultimately a by-product, or side-effect, of modern civilisation: a consequence of advancing science, the accelerated speed of life, the global diffusion of knowledge, mass marketing, consumer choice and individual freedom. As such it was curiously symmetrical with the enervated state of neurasthenia that many of its sufferers had first turned to morphine to cure. George Miller Beard considered that the dangers of drugs had 'greatly extended and multiplied with the progress of civilisation, and especially in modern times'.[18] He grouped it together with other diseases of 'overcivilisation' – suicide, pre-marital sex, homosexuality – as an unwanted but inevitable by-product of the inventions he saw as the core drivers of modernity: the printing press, the steam engine and the telegraph.

As the number of cases grew and medical specialists proliferated, it became clear that this disease of civilisation had two primary classes of victim. One was wealthy private patients; the other was doctors themselves. In 1883 the American physician J.B. Mattison claimed that his colleagues formed the largest group of habitués, or morphine addicts, in the nation, and might include as many as a third of all physicians, surgeons, dentists and nurses. By the early twentieth century, the leading German pharmacologist Louis Lewin estimated, around 40 per cent of addicts were doctors, and a further 10 per cent doctors' wives.[19] By 1886, cocaine was beginning to be considered by some doctors as comparable in its dangers to morphine: as the *British Medical Journal* put it, 'we have already found out that this sweet rose of our therapeutic bouquet has a bitter thorn'. It was particularly prevalent among anaesthetists and surgeons, the specialists who worked with it on a daily basis. Its addicts did not suffer the same agonies of withdrawal as morphinists, but it could be more insidious: the Irish addiction specialist Conolly Norman wrote in the *Journal of Mental Science* in 1892 that 'cocaine is more seductive than morphia; it fastens on its victim more rapidly, and its hold is at least as tight'.[20]

The dangers of morphine and cocaine were considerably magnified by the modern method of administration that they shared. In the first sentence of his book, Eduard Levinstein had noted that the new condition was intimately connected to 'Pravaz's method' of subcutaneous injection, named for the French surgeon Charles Pravaz who, in 1853, designed and commissioned a hypodermic syringe, made entirely of silver, to dispense 1 cubic centimetre of liquid through a fine, hollow needle.[21] The US Civil War and the Crimean War in Europe had entrenched morphine injections in battlefield medicine, and during the 1870s the Pravaz, together with morphine vials and tablets, became an indispensable element of the doctor's bag.

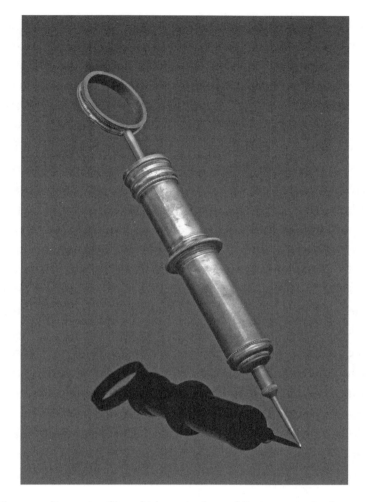

*5. The new diagnosis of 'morbid craving' or addiction was closely associated
with the fast-growing use of the hypodermic needle.*

Following on from the adoption of surgical anaesthesia in the
1840s, subcutaneous injection marked a triumphant new era in medicine. Throughout human history, pain had been a universal constant:
everyone who had ever lived had expected to suffer unbearable pain at
some point in their lives, and often at their death. With the combination of morphine and the hypodermic, however, previously intractable

pain could be removed entirely, more or less instantly. The growing awareness of the danger this represented can be traced through one of the first texts to teach the new technique, the *Manual of Hypodermic Medication* by the Philadelphia physician Philips Bartholow. In its first editions of 1869 and 1873, Bartholow mentioned the morphia habit only briefly; by 1879 it warranted a full chapter, and the admonition that 'the introduction of the hypodermic syringe has placed into the hands of man a means of intoxication more seductive than any which has heretofore contributed to his craving for narcotic stimulation'.[22]

With the new drugs and the new method, pain could be instantly eliminated and pleasure spontaneously created. The two sensations had been yoked together in Humphry Davy's nitrous oxide experiments, in which he proposed that they were opposite forms of stimulus: the gas, he theorised, eliminated pain by temporarily flooding the same nerves with pleasure. Thomas Beddoes had hailed the discovery as the dawn of a new scientific age in which man would 'come to rule over the causes of pain and pleasure, with a dominion as absolute as that which at present he exercises over domestic animals'.[23] With morphine, cocaine and the Pravaz, Beddoes's prophecy had come to pass, but it proved more contentious than he had imagined. Euphoria in unlimited doses turned out to be a medical problem in its own right. Eduard Levinstein, in his *Morbid Craving for Morphia*, connected it to the problem of addiction by suggesting that the pains of withdrawal were a mirror image of the drug's initial pleasures: 'It follows that the opposite of this euphoria, this high degree of self-feeling, is a deep state of depression.'[24] The control of pain, the great discovery of modern medicine, trailed an ominous shadow.

*   *   *

Other clinicians echoed Dr Erlenmeyer's alarm at the new scourge of humanity, and the diagnosis of 'cocainomania' soon joined that of morphinomania. At the same time, pharmaceutical companies were vigorously promoting cocaine as the miracle of the age. In their promotional brochure of 1885 Parke, Davis celebrated it as a 'universal panacea' that can 'supply the place of food, make the coward brave, the silent eloquent'.[25] It quoted neurologists who had been using it successfully in the treatment of nervous disorders, including one who recommended it for the treatment of morning sickness in pregnancy. Its value was, Parke, Davis claimed, also being recognised in the treatment of alcohol and opiate habits. One of the authorities quoted was 'Dr Sigmund Freud of Vienna', who had cured a case of severe morphine addiction in ten days with the cocaine method and 'is of the opinion that a direct antagonism exists between morphine and cocaine'. Freud's opinion was echoed by his patient Dr Fleischl, who added that cocaine was so effective in treating the alcohol habit 'that inebriate asylums can now be entirely dispensed with'.[26]

In the summer of 1885 Parke, Davis approached Freud to evaluate their product, as he had done for Merck. He obliged, deploying his dynamometer once more to measure its effects:

> I have examined cocaine muriaticum [hydrochloride] produced by Parke Davis for its physiological effects and can state that it is fully equal in effect to the Merck preparation of the same name. When taken internally it produces the characteristic cocaine euphoria. Increases in muscular strength were measured with the dynamometer after equal doses of Parke and Merck cocaine, and they were found to be the same. Parke's cocaine, when applied in 2% solution, anaesthetises the cornea and conjunctiva of the eye equally to the Merck product.[27]

Freud was now leveraging his professional expertise to advocate for a commercial pharmacy product, a process that was shaping its applications and the way its effects were understood. Its euphoriant qualities may have been a source of anxiety for addiction specialists, but for pharmacists and general practitioners they were a supreme selling point. The 'feeling of contentment and well-being' it generated was, one doctor reported, a sovereign remedy for depression – 'the blues, in other words'.[28] Other physicians testified in Parke, Davis's journal to its powerful benefits for 'the nervous and depressed' and in cases of melancholia. One praised 'the almost mathematical precision of the effect' on restoring exhausted nerves.[29] Euphoria, for those selling it, was not a cause or a symptom of neurasthenia but a remedy for it, precisely as Freud had hoped when he began his investigations.

In the fall of 1885, with a travel grant awarded him by the University of Vienna on the strength of Ernst Brücke's warm recommendation, Freud took up residence in Paris to study at the Salpêtrière Hospital under the world-famous neurologist Jean-Martin Charcot. It was an exciting but daunting assignment: he was thrown into a world of lectures, hospital visits and social engagements with the luminaries of his intended profession, in a dazzling and intimidating city, speaking very little French and attempting to survive on a poverty-level income. He found himself using cocaine more regularly, in ways that embraced both medicine and pleasure. After the austerity of Vienna General Hospital and the tight biomedical focus of Brücke's tutelage, he was overwhelmed equally by Charcot's charisma, his *bon vivant* lifestyle and his close attention to the patient as a living, embodied, conflicted subject. His letters to Martha describe Charcot as 'like a worldly priest from whom one expects a ready wit and an appreciation of good living'.[30] He confided to her that he was using cocaine to calm his nerves and overcome his shyness and poor language skills during formal dinners, and as a remedy for the neurasthenia 'produced by the toils, the worries, the excitements of these last years'.[31]

* * *

For the general public, the new drugs transforming medicine were mostly encountered in one of the commercial wonders of the age, the modern pharmacy. Until mid-century, chemists' shops had typically been dimly lit stalls, their shelves lined with bottles of 'simples': plain powders and liquid preparations of drugs and chemicals, poured to order into twists of paper or small bottles. By the 1880s, however, metropolitan flagship stores in particular had become consumer palaces and temples to medical progress. Their mirrored and plate glass entrances were portals to interiors illuminated by gas or electricity in which a rainbow of vividly coloured pills, tablets and lozenges was arrayed across, behind and beneath the gleaming service counters. Their distinctive motif, still to be found today, was the carboy: a decanter-shaped glass vial, several often placed in regimented rows in the storefront or across the top shelves, filled with luridly coloured water and backlit to glow like precious gems. The aniline dyes that provided their colour spoke to the bright promise of modern industrial chemistry, while the shape of the carboy also suggested the Oriental souk with its exotic wares of scents, potions and elixirs of dissolved pearls or rubies. Customers' expectations of the new drugs were coloured by their setting, which framed them equally as miracles of modern science and mysterious, quasi-magical potions.

'Simples', including pure cocaine, were available from the new pharmacies, but the big sellers were proprietary blends, boldly branded and advertised. Pharmaceutical manufacturers such as Parke, Davis and Merck supplied what were known as 'ethical' products, manufactured to clean and modern laboratory standards, with doses of known strength and purity accurately labelled. Beyond these was a larger market of 'patent' medicines, a misnomer since their ingredients were usually not listed, let alone patented. They were typically bulked out with inert fillers such as soap, turpentine and wax – one leading brand, Holloway's Pills and Ointments, was shown to contain nothing else[32] – but by the 1880s many included narcotics and stimulants such as opium, morphine and cocaine. These were often branded with a folksy figurehead, such as 'Mrs Winslow's Soothing Syrup' or 'Godfrey's Cordial', and their bitter alkaloids offset with sickly sweet flavoured syrups. Whether these medicines 'worked' was essentially in the mind of the consumer, who would likely feel better after taking them. Some

6. In the late nineteenth century, pharmacies were a wonder of the modern consumer age.

brands were potent cocktails of the broad-spectrum euphoriants that would be classed as 'drugs' in the decades to come: a staple of British pharmacies, for example, 'Dr J. Collis Browne's Chlorodyne', contained chloroform, ether, morphine and cannabis. On the label, it was advertised as 'a medicine chest in itself'.

The effects of the new pharmacy drugs, and the promises and threats they represented, were a staple of newspaper columns and grist to the fast-turning mill of popular fiction, which established its popularity in American magazines such as *Lippincott's* and *Scribner's* and by the 1890s had spread to Britain, ushering in what became known as the 'Age of the Storytellers'.[33] The question of stimulants was examined with a quizzical eye by H.G. Wells in his short story 'The New Accelerator', which appeared alongside the latest episode of Conan

Doyle's hit Sherlock Holmes story *The Hound of the Baskervilles* in *The Strand Magazine* in December 1901, an issue that sold around half a million copies. Like most drug stories of the era, it revolved around a self-experimenting scientist. The renowned Professor Gibberne, an expert in the field of 'soporifics, sedatives and anaesthetics', is 'seeking an all-round nervous stimulant to bring languid people up to the stresses of these pushful days' and stumbles upon 'something to revolutionise human life'.[34] The narrator, a neighbour of Gibberne, accepts a dose of the compound the professor has christened the New Accelerator. Shortly after swallowing a solution in water, he notices that the external world has slowed almost to a standstill, and realises that his own actions are taking place in impossibly tiny increments of normal time.

The pair wander around the genteel British seaside town of Folkestone, too rapidly to be perceptible to the passers-by, whom they observe in grotesque, slow-motion detail. The experience is exhilarating and unsettling in equal measure. It announces a future in which people will be able to select the speed at which they move through time, capable of living a day in a fraction of a second. Gibberne enthuses on the possibilities this will open up: 'the convenience of securing a long, uninterrupted spell of work in the midst of a day full of engagements cannot be exaggerated'.[35] He is beginning work on a Retarder, a drug to dilute the Accelerator's alarming potency and perhaps achieve the opposite effect of allowing a long and tedious passage of time to pass in a subjective instant. In the meantime, however,

> Its appearance on the market in a convenient, controllable and assimilable form is a matter of the next few months. It will be obtainable of all chemists and druggists, in small green bottles, at a high but, considering its extraordinary qualities, a no means excessive price. Gibberne's Nervous Accelerator it will be called, and he hopes to be able to supply it in three strengths: one in 200, one in 900, and one in 2000, distinguished by yellow, pink and white labels respectively.[36]

Wells, at this stage a science journalist as well as fiction writer, followed medical and pharmaceutical discoveries closely and was intrigued by the prospect of a stimulant that could boost the energy available to the nervous system. His 'New Accelerator' was a fantastical extrapolation of cocaine, along with the more drastic and improbable glandular and electrical stimulants. It glimpsed the thrilling and terrifying prospect of a world in which drugs might loosen the bonds between individuals to the point where they came to inhabit not only private and subjective mental worlds, but quite different realities. It reflected a modern marketplace in which mind-changing drugs progressed effortlessly from dazzling discovery to mass-market commodity, quietly rewriting the human condition in the process. All this was proceeding with no oversight or effective means of control, and the story concludes with Gibberne's airy, *laissez-faire* dismissal:

> 'Like all potent preparations, it will be liable to abuse. We have, however, discussed this aspect of the question very thoroughly, and we have decided that this is purely a matter of medical jurisprudence and altogether outside our province. We shall manufacture and sell the Accelerator, and as for the consequences – we shall see.'[37]

In 1887 the *British Medical Journal* pronounced that 'an undeniable reaction against the extravagant pretensions announced on behalf of this drug has already set in'.[38] Yet the judgement was by no means settled. For some, the dangers of cocaine were an indictment of consumerism, for others of the stresses of modern life; for others still, they exposed the existence of a minority of moral degenerates. The opinions of doctors, pharmacists and the public at large all varied widely. There were still many physicians who found cocaine useful, even indispensable, for applications ranging from ophthalmology to depression, local anaesthesia to gastric complaints, and believed its dangers to be overblown. Injecting clearly brought its own risks – septicaemia, addiction, nervous collapse – but there was little evidence that everyday consumers of coca wines and lozenges were experiencing anything worse than mild symptomatic relief for their toothaches, asthma or 'the blues'.

Freud's next contribution to the debate, his 1887 paper 'Craving for and Fear of Cocaine', showed how hard it had become to speak across this divide. Cocaine addiction, he argued, was not a disease but merely a symptom of other mental disorders. He rejected Erlenmeyer's characterisation of the drug as the 'third scourge' of humanity' as 'pathetic', arguing that 'all reports of addiction to cocaine and deterioration resulting from it refer to morphine addicts', whose abuse of it was part of a previously established pattern of chaotic self-harm.[39] To make this case, he was obliged to retreat from his recommendation that it was a remedy for morphine addiction, a claim he now attributed to Parke, Davis, whose advertisement of it had brought it 'to the general attention of physicians – and also, unfortunately, of morphine addicts'.[40]

There were, he acknowledged, rare examples of cocainomania, but these were more correctly attributed to the hypodermic needle, a device that offered immediate gratification but only at the risk of rapidly escalating tolerance that pushed cocaine use to toxic levels, causing physical agitation, delirium and persecution mania. Freud glossed over the fact that he had himself recommended injection in print in 1885, and that Fleischl had been using this method, if not with Freud's explicit approval at least under his watch, at the time that Freud was using his anonymous case as a success story. In rescuing cocaine from the 'slanders' levelled against it, he was pushed into self-contradiction and falsehood – or at the very least, in the exculpatory reading of his first biographer Ernest Jones, to unconsciously erasing the facts that undermined his position.[41]

Erlenmeyer's denunciation of cocaine was winning over the risk-averse middle ground of medical opinion, and with hindsight marked the end of Freud's personal ambitions for the drug. But other medical figures were unwilling to declare it a scourge of humanity. Freud concluded his paper by quoting at length the self-experimental report of William Hammond, a truculent advocate for cocaine who had no intention of changing his opinion. Hammond was among the most distinguished doctors in the United States: he had served as surgeon-general of the Union Army during the Civil War, and had worked energetically to professionalise wartime medicine. He was controversially removed from the post for refusing to administer the mercury-containing emetic calomel to sick personnel: he claimed it was neither safe nor effective, and was later proved right. He demanded a court-martial, which dismissed him on vague grounds of 'irregularities', but he went on to forge a successful career as a neurologist, becoming professor of nervous diseases at New York University in 1874.

Hammond approached cocaine as a libertarian, a rationalist who tirelessly debunked the claims of spiritualists and quacks, and an advocate of new chemical treatments such as lithium for mania. He gave it a far more thorough trial than Freud ever had:

> I began by injecting a grain of the substance under the skin of the forearm, the operation being performed at 8pm. The first effect ensued in about five minutes, and consisted of a pleasant thrill which seemed to pass through the whole body . . . On feeling the pulse five minutes after making the injection, it was found to be 94, while immediately before the injection it was only 82. With these physical phenomena there was a sense of exhilaration and an increase of mental activity that were well marked, and not unlike in character those that ordinarily follow a glass or two of champagne. I was writing at the time, and found that my thoughts flowed with increased freedom and were unusually well expressed.[42]

A couple of days later Hammond doubled the dose, bringing it up to 130mg. He noted the same physical sensations and a 'great desire to write', which yielded a text 'that was entirely coherent, logical, and as good if not better in general character as anything I had previously written'. On subsequent evenings he increased the dose further until he was injecting 12 grains, the best part of a gram, at which level 'the action of the heart was increased, was irregular in rhythm and force to such an extent that I was apprehensive of serious results'.[43] This did not deter him from making a final heroic experiment with 18 grains, well over a gram, taken in four injections within five minutes of each other. 'In this instance', he recorded:

> I felt that my mind was passing beyond my control, and that I was becoming an irresponsible agent . . . I lost consciousness, I think, of all my acts within, I think, half an hour after finishing the administration of the dose. Probably, however, other moods intervened, for the next day when I came downstairs, three hours after my usual time, I found the floor of my library strewn with encyclopaedias, dictionaries and other books of reference, and one or two chairs overturned.[44]

'Certainly in this case,' Hammond concluded, 'I came very near taking a fatal dose and I would not advise anyone to repeat the experiment.'[45] Yet he noted no morbid cravings or withdrawal symptoms, a point he emphasised in a lecture to the New York Neurological Society in 1886 entitled 'The So-Called Cocaine Habit'. He told his audience that the cocaine habit was in reality 'similar to the tea and coffee habit, and unlike the opium habit':[46] it amounted to a 'pleasurable mental exhilaration' that created no metabolic dependency. It required merely will-power to resist, 'and nothing like as much as stopping alcohol or tobacco'. Hammond did not dispute Erlenmeyer's testimony that cocaine addicts existed, just as coffee addicts did, and he emphasised that cocaine grafted onto a morphine habit was 'an exceedingly bad combination'.[47] In his view, however, cocaine was no scourge of humanity but simply another addition to the long list of substances that were harmful at inappropriately high doses. He continued to enjoy a glass of coca wine regularly, 'at the close of his day's duties'.[48]

The response to Hammond's lecture revealed a profound lack of consensus among New York's physicians. One Brooklyn doctor had witnessed seven cases of cocaine addiction – five physicians and two pharmacists – and firmly believed it caused physiological damage 'more unfavourable even than morphine'.[49] Others, however, found it effective in cases of depression, or by injection for the relief of sciatica, as an anaesthetic in haemorrhoid operations or in the treatment of conjunctivitis. One felt strongly that the alarm about its addictive properties was creating 'prejudice against a most useful remedy'.[50] Freud had stressed in 'Über Coca' that individual responses to cocaine varied widely: from anxiety to euphoria, increased energy to dizziness or nausea. He had also noted a wide variation in dose response: 'I have found not a few who remained unaffected by 5cg [50mg], which for me and others is an effective dose':[51] this was one of the reasons he had performed the majority of his dynamometer tests on himself. In particular, 'the subjective phenomena after the ingestion of coca differ from person to person, and only a few persons experience, like myself, a pure euphoria'.[52]

A settled, orthodox medical opinion was, it seemed, elusive where mind-altering drugs were concerned. The views of physicians were strongly coloured by personal experience, which varied widely; experience, in turn, tended to confirm the prejudices that preceded it. For every William Hammond there was a case such as William Halsted, America's most brilliant surgeon, resident at New York's Bellevue hospital. Halsted was an early adopter of cocaine, having visited Vienna in 1880 and read Köller's paper on cocaine anaesthesia on publication. By the end of 1884 he was experimenting with it in his surgery, numbing nerves and muscles with precision and testing which surgical procedures it could be used for. In the evenings he injected his assistants and medical students in different sites, and they took to sniffing it before theatre outings or adjourning to Halsted's opulent town house on 25th Street.[53]

By the spring of 1885 his habit had escalated out of control and his professional life unravelled. He missed lectures, delegated operations to his colleagues and at the April meeting of the New York Surgical

Society, where he normally shone with wit and expertise, he cut a shambling, confused figure. By this time he had walked out of Bellevue and was holed up in his house with large quantities of cocaine. In company he was domineering and excitable, steamrollering visitors with monologues 'about everything under the sun from the transit of Venus to gonococci'.[54] His medical colleagues intervened, dispatching him on a cruise to the Caribbean with only enough cocaine to taper his doses down to zero. Halsted lay in his bunk, obsessively calculating how long his supply would last, until one night he broke into the captain's medicine chest. By the time the ship made its way back to Florida he was exhausted, insomniac, paranoid and tormented by aching muscles that had been tightly clenched for weeks. On his return to New York he admitted himself, under pressure from friends and colleagues, to a residential mental asylum in Providence. After several months' further detoxification in a sanatorium, he recovered to become the first professor of surgery at Johns Hopkins University.

There was no consensus, at the time or subsequently, about why Halsted had succumbed so disastrously to cocaine, or indeed whether he ever truly recovered from his addiction. His first biography, published in 1930, passed swiftly and discreetly over the episode, characterising it as an accident that befell many early cocaine users who were at that time 'quite innocent of any knowledge of its habit-forming character'; the narrative focused on how he had 'conquered it through superhuman strength and determination and came back to a splendid life of achievement'.[55] A later account by his close friend, the neurosurgeon Harvey Cushing, probed more deeply, noting that before he took the drug Halsted had been 'a rigorous, rather showy, didactic, bustling individual'. Cocaine magnified all these tendencies, and 'the truth of the matter is that he never conquered it'. Rather than destroying his skills, however, Cushing suggested that cocaine had increased his fastidiousness, making him even more obsessive about detail and hygiene, and 'the whole Halsted school of surgery which I have called a School for Safety in surgery may have been due to this drug addiction'.[56]

* * *

Stories even more alarming than Halsted's bubbled up regularly from the medical literature. The German toxicologist Louis Lewin, who by the 1880s had established himself as an international authority on mind-altering plants and drugs, wrote that one of his patients, suffering from facial neuralgia, had been resorting to morphine for pain relief until he was introduced to cocaine. He was soon using over a gram a day, soaked into cotton-wool plugs and inserted between his teeth:

> The unfortunate man's own words were as follows: 'With regard to the action on my personality, I can honestly declare that the past five years can be counted among the happiest of my life, and I owe this primarily to cocaine. Nothing can refute this plain fact.' His letter of twelve pages terminates with these words: 'Time is necessary to bring my conception of the world to a point which is founded on this sentence: "God is a substance!"'[57]

As addiction became widespread among doctors, dentists, surgeons and their families, the risk to the unsupervised general population became more alarming. The free availability of drugs such as cocaine from pharmacies was accompanied by a thriving market for home-doctoring manuals and other guides encouraging the public to avoid doctors' fees by learning the rudiments of pharmacology themselves. Apothecaries and pharmacies had always fed a subculture of self-doctoring among the public – 'quacking oneself', in the old eighteenth-century term – and the new drugs allowed individuals to experiment with novel moods and perceptions, on a spectrum that extended from mild and manageable euphoria to excess, compulsion and mental collapse.

Then as now, medical evidence and opinion was skewed towards the cases witnessed by doctors, which were those where self-experiments had gone disastrously wrong. It is hard to estimate how many members of the public experimented safely and productively, as there were many disincentives to publicising one's personal drug use; but one unusually well-documented example makes it clear that some members of the general public developed an understanding of drugs that was considerably more sophisticated than most doctors'. Between 1895 and 1914 the British engineer James Lee pursued a career in construction and mining projects in the colonies, mostly across south and east Asia. In 1935, after he retired, he published his memoirs under the title *Underworld of the East: Being Eighteen Years' Actual Experience of the Underworlds, Drug Haunts and Jungles of India, China and the Malay Peninsula*. Lee's story offers a quite different perspective from the medical literature, a rational and practical approach to using drugs and managing their pleasures and pains:

> The life of a drug taker can be a happy one; far surpassing any other, or it can be one of suffering and misery; it depends on the user's knowledge. The most interesting period will only be reached after many years, and then only if perfect health has been retained.[58]

Lee's experiences may have been exceptional but his background was not. He was one of the new generation of British working men, state-educated in the wake of the 1870 Education Act, with access to a network of scientific societies and lending libraries, and eager to expand his personal horizons. Born to an iron merchant in the industrial north-east of England, he became an engineer's apprentice in Sheffield and Teesside at seventeen before moving to London to take up a post as a teaching assistant in a school of mechanics. At the age of twenty-one, he applied for an advertised vacancy as a mining foreman in Assam. He was, he wrote, becoming bored with life in England: 'There was too much sameness about it; a place where there is little real freedom, and where one had to do just as the next fellow did. To wear the same kind of clothes with a collar and tie, and talk about football and horse-racing, or be considered no sport.'[59]

Lee's drug career began during his first assignment in a remote rural district of Assam in north-east India, where he soon found himself suffering from malarial chills and fevers. The local doctor promptly gave him an injection of morphine that left him 'simply purring with content', and sent him home with a syringe kit and a tube of morphine tablets. Before long Lee found himself 'looking forward to the afternoon when the day's work was over, and I could take a larger dose and lay dreaming rosy dreams'.[60] The doctor taught him the basics of safe injecting; he learned to boil and sterilise his needles, and to recognise the signs of sepsis. After a few weeks he noticed that his tolerance to morphine was increasing and decided to give up the habit, but this proved easier said than done. As he reduced his dose he was tormented by cramps, insomnia and a 'horrible feeling of depression and gloom'.[61] He returned to the doctor, who told him, 'Sir, morphia is a very strange medicine, it is both Heaven and Hell. It is very difficult to give up, but it can be done.'[62] With that, he injected Lee with half a grain of cocaine.

Lee found the new drug 'stimulating and exhilarating, producing a feeling of well-being; of joy and good spirits', but after a while it gave him insomnia.[63] This time the doctor's remedy was to invite him to his home for a few pipes of opium in the evenings, and sleep was restored. Lee, however, decided to rid himself of his inadvertently acquired habit. Addiction, in most drug literature the terminal destination, was in Lee's case only the beginning of the journey: 'I now started to use drugs scientifically,' he wrote.[64] Used immoderately, he recognised, cocaine would eventually reduce the user to a skeletal nervous wreck, just as morphine would become a cul-de-sac that consumed waking existence entirely; but the combination, used judiciously, rescued him from both these destructive tendencies. He developed a regime in which he alternated the two drugs, tapering one and replacing it with the other in carefully calibrated doses, and developed a detox regime that he spells out in detail, grain by grain. 'These two drugs', he noted, as Ernst Fleischl had done, 'are in a certain way antidote to each other': each could be used to reduce the cravings for the other, and both gradually watered down.

I admit that at the end I had a little craving, but it was nothing really, and I was getting freer of it every day. Still, I decided that the system was not perfect, and I meant to continue experimenting and searching, until I found a cure which was fool proof and easy.[65]

He took to using this cure on his return trips to England: the six weeks of the voyage was just enough to allow him to step off the ship at Portsmouth healthy and drug-free. The broader insight to which this led him was that 'one drug alone spells disaster': only with combinations of different drugs could their desirable effects be maintained.[66] As he pursued his self-described 'hobby' over the years, he discovered that periods of total abstinence were also necessary to allow full physical recovery.

One particularly valuable aspect of Lee's narrative is that he appears entirely ignorant of the debates over cocaine. He is clearly unfamiliar with Freud and the controversy over the morphine-cocaine cure, and appears not to have heard of Thomas De Quincey, Charles Baudelaire or any of the popular drug literature that insisted that the pains of drugs would always win out over their pleasures. He avoided newspapers and the company of his fellow Europeans, whom he rarely encountered in the remote corners of India and, later, the Malay archipelago and China. Instead, he soaked up information from his Hindu doctor, and from the local populations on his travels, whose traditional haunts he sought out and frequented. He married a local woman called Mulki, who had fled an arranged marriage in the Central Provinces for a life of hard labour in in the mines of Assam. But his primary source of knowledge was his self-experiments. Over the years he developed a healthy distrust of Western medical opinion, which usually turned out to be founded on much less knowledge than his own. On one of his voyages back to Europe, his carefully laid detox plans were scuppered by his cabin-mate, a young medical student:

Soon he discovered that I was using drugs, and he gave me a lecture on the terrible consequences of the habit. I asked him if he had ever taken any himself, and he confessed he had not, and that he was going on what he had heard.

Shortly afterward I missed my syringe.[67]

'In morphia are combined a blessing and a curse', Louis Lewin wrote, echoing Lee's Indian doctor, though for Lewin the difference was that 'if it is dispensed by the hand of a physician its power is divine'.[68] But the outcome, it often seemed, depended not so much on medical expertise as on the character and situation of the user. The difference between a James Lee and an Ernst Fleischl was bound up in how they had acquired their habits and what was at stake in quitting them. Lee's dependency was accidental, and overcoming it was a matter of pride and self-respect; Fleischl was in constant and agonising pain, to which abstaining from morphine and cocaine immediately returned him. Faced with this prospect, he acted in ways that were completely outside his normal character: concealing his habit, lying to his friends, using cocaine to self-destructive excess. To friends and medical authorities alike, this appeared as a dramatic transformation of personality, as if the subject had been taken over by a second self, an alien or demonic force. When Harvey Cushing wrote his memoirs of William Halsted in 1931, he described how cocaine had transformed him from a 'brilliant, rapid, spectacular operator' to a monster of egotism and self-indulgence. By this time there was a well-established shorthand for this transformation. Halsted had become, in Cushing's words, 'a Jekyll and Hyde character'.[69]

* * *

The Strange Case of Dr Jekyll and Mr Hyde emerged in 1886, just as Sigmund Freud was publishing his papers on cocaine, and quickly established itself as the defining metaphor for drug-induced personality change. It can be read in innumerable ways: together with Frankenstein or Prometheus, as a story of scientific hubris; as a parable

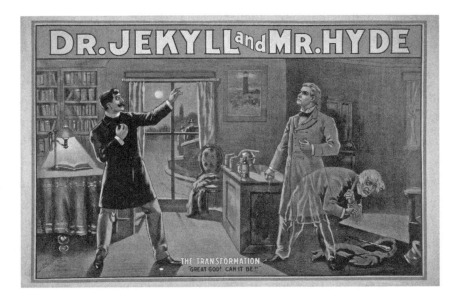

7. The Strange Case of Dr Jekyll and Mr Hyde *(1886) became the classic parable of self-experimentation and the dangers of personality-altering drugs.*

of the struggle between good and evil for the soul of man; as a psycho-geography of the modern city, its genteel uptown residences secretly connecting to the dark underbelly of its tenements and slums. At its most literal level, however, it is the tale of the disaster that a brilliant medical researcher brings upon himself by self-experimenting with an intoxicating white powder.

Robert Louis Stevenson, a chronic sufferer from nervous exhaustion, wrote the final version of his novel in three days and nights, sustained by a cluttered regime of tonics, pick-me-ups and nervous stimulants that included coca wine.[70] But whether or not it was a direct inspiration, Dr Jekyll's first reaction to his potion is a pure euphoria almost like a paean from the pen of Paolo Mantegazza:

There was something strange in my sensations, something indescribably new, and from its very novelty, incredibly sweet. I felt younger, lighter, happier in body; within I was conscious of a heady recklessness, a current of disordered sensual images running like a mill race in my fancy . . .[71]

This euphoria, however, was anything but normal or healthy. It lacked the innocence of true happiness: it was something 'more wicked, tenfold more wicked', a suppressed shadow self that had taken possession of its previously sober host. Hyde was still conscious of his identity as Dr Jekyll, yet as he cast off his sober scientific self 'the thought, in that moment, braced and delighted me like wine'.[72] The source of the evil, it transpires, was his habit over the years of suppressing his desire for pleasure, the gratification of which he had found 'hard to reconcile with my imperious desire to carry my head high, and wear a more than commonly grave countenance before the public'.[73] His had been 'a life of effort, virtue and control', and with the potion he had unwittingly released an atavistic urge, a repressed self that seized its chance to take possession of him.

The story originally came to Stevenson in the form of a nightmare that he decided, on waking, would make 'a fine bogy tale'.[74] It was not intended as a moral tract, and it offered no answer to the question of whether Jekyll had been right to 'conceal his pleasures' or what he might have done instead.[75] Like the cocaine user, his alter ego had greater reserves of energy and a sharper intelligence than his normal self, and Jekyll found something uncanny in the preternatural brightness that met him in the mirror. The glimpse of 'two natures that contended in the field of my consciousness', as Jekyll puts it, confronted him with the realisation that the base, instinctive self was bound to gain the upper hand. The potion, in this sense, warned of the consequences of a society under the influence of ever stranger and more powerful drugs, all giving uninhibited rein to the 'primitive duality of man'.[76] When Jekyll sees the monstrous face in the glass:

I was conscious of no repugnance, rather a leap of welcome. This, too, was myself. It seemed natural and human. In my eyes it bore a livelier image of the spirit, it seemed more express and single, than the imperfect and divided countenance, I had been hitherto accustomed to call mine.

Jekyll and Hyde was a popular sensation, and a pervasive influence on the many gothic fictions of the *fin-de-siècle* that were hung around the ambivalent figure of the drug experimenter. Often the protagonist-victim was a 'brain-worker', a high-risk category for neurasthenia and the risky overuse of stimulant drugs: as the leading French physiologist Charles Richet described them in the *Popular Science Monthly*, 'poisons of the intelligence'.[77] An exceptional mind was no defence against them; the keener the intellect, the more liable it was to turn destructively on itself under their influence. Enervated scholars had long run the risk, in fiction at least, of being driven out of their wits by sleepless nights under the influence of coffee or tea, but the arrival of newly potent stimulants allowed authors to stretch the consequences into the domain of the supernatural. The bestselling fantasy author George Griffiths, for example, in his short story 'Genius for a Year' (1899), imagined an author tormented by suspicions of his artistic mediocrity and driven to experiment with some hashish mailed to him by his brother from Calcutta. After taking the first dose, he seemed to be observing himself from a distance as he, or his alter ego, wrote page after page effortlessly, eyes burning 'with a fierce light that might have been either insanity or genius'.

As the drug wore off, 'his two beings seemed to fuse together and become one', and he fell into a dreamless sleep.[78] When he awoke and read his nocturnal scribblings, he recognised the work as a masterpiece beyond anything he had previously written. For a year, genius possessed him, but as he read over his nearly completed novel, 'he saw that it was beautiful, but it was utterly strange to him. Who had written it?' In a flash of illumination, he swallowed all his remaining hashish pills, and 'visions of chaotic splendour chased each other in headlong haste through the death-dance of his expiring senses'. He was found dead, his writing table 'strewn with pages filled with the most hideous nonsense', ending in an 'unintelligible scrawl'.[79]

A similar fate befell the protagonist in Arthur Machen's 'The Novel of the White Powder' (1895), in which 'an innocent-looking white powder, of which a little was dissolved in a glass of cold water', gradually transformed Francis, an anxious and bookish young man, into 'a lover of pleasure, a careless and merry idler of western pavements, a hunter out of snug restaurants, and a fine critic of fantastic dancing'.[80] His friends, who had blamed his long hours of study for 'a little mischief in the nervous system', were initially delighted, but his sister became concerned when she noticed a sinister mark on his skin, between thumb and forefinger. Soon Francis confined himself to his room, refusing visitors; when his friends finally forced the door, he was revealed in an advanced state of ghastly physical corruption.

Machen was drawing on Oscar Wilde's *The Picture of Dorian Gray* (1890) as much as *Jekyll and Hyde*, and the visceral and hallucinatory dénouement looked forward to the twentieth-century's pulp horror, particularly that of his great admirer H.P. Lovecraft. The stimulant white powder was revealed to have been a toxic batch that had been decomposing on the damp shelves of the local pharmacy for decades, but its true danger was that it granted unearned pleasure, a corrupting influence that negated the sense of shame and – a distinctive refrain in Machen's horror fiction – self-disgust. Francis's final dissolution in his bedroom reflects *fin-de-siècle* theories of nervous disease and degeneration, and connects the gratification from a drug such as cocaine with the 'solitary vice' of self-pollution: an empty and mechanical pleasure that, far from increasing the reserves of nervous energy in the system, drains them beyond repair.

The most extended engagement with cocaine in the fiction of the era was, of course, in the stories of Sherlock Holmes, and Arthur Conan Doyle's handling of the theme is an eloquent illustration of the shift in public opinion through the 1890s. The great detective's early adventures are the ones most liberally spiked with drug references. Initially, Holmes's detective work seems almost to play second fiddle to his self-experiments: in the first published short story, 'A Scandal in Bohemia', we hear of him 'alternating from week to week between cocaine and ambition', until he 'had risen out of his drug-created dreams, and was hot on the scent of some new problem'.[81] In 'The Five Orange Pips', which appeared in *The Strand Magazine* of November 1891, Dr Watson describes him as 'a self-poisoner by cocaine and tobacco'.[82] But it was the appearance of *The Sign of Four* in 1890, the year *The Strand* established itself as the leading purveyor of illustrated fiction and Sherlock Holmes as its undisputed star, that his drug habit was most memorably established. In its celebrated opening lines:

> Sherlock Holmes took his bottle from the corner of the mantelpiece, and his hypodermic syringe from its neat morocco case. With his long, white, nervous fingers he adjusted the delicate needle, and rolled back his left shirt-cuff. For some little time his eyes rested thoughtfully upon the sinewy arm and wrist, all dotted and scarred with innumerable puncture-marks. Finally, he thrust the sharp point home, pressed down the tiny piston, and sank back into the velvet-lined armchair with a long sigh of satisfaction.[83]

'Which is it today,' Dr Watson asks, 'morphine or cocaine?' This is the only occasion on which Holmes is described as a morphine user: in subsequent stories, his habit is confined to cocaine. It may be that Doyle decided that injecting morphine seemed too much like a medical condition, whereas cocaine could be interpreted as a vice or eccentricity of character. Still a practising doctor at this point, Doyle was familiar with both drugs: indeed, in 1891 he spent a few months studying ophthalmology in Vienna, where Köller's discovery of cocaine anaesthesia had transformed eye surgery. The tenor of conservative medical opinion was captured authentically in Dr Watson's disapproving comments on Holmes's habit – 'it is a pathological and morbid process, which involves increased tissue-change, and may at last leave a permanent weakness' – a serious-sounding diagnosis that is nonetheless so vague that it could refer to anything from masturbation to keeping late hours or living in the tropics. It succeeds, however, in prompting Holmes to the plainest and most enduring statement of his motivations. 'My mind,' he replies,

> rebels at stagnation. Give me problems, give me work, give me the most abstruse cryptogram, or the most intricate analysis, and I am in my own proper atmosphere. I can dispense then with artificial stimulants. But I abhor the dull routine of existence. I crave for mental exaltation. That is why I have chosen my own particular profession, or rather created it, for I am the only one in the world.[84]

Here Doyle places stimulant drugs at the centre of both Holmes's professional world and his inner life. He is a brain-worker, craving stimulation but also indulging a neurasthenic and self-destructive streak. He is described elsewhere as a victim of 'black moods', constantly swinging from obsessive work to periods of depression – 'in the dumps' – where he speaks to no-one for days on end.[85] At the same time, he is a committed aesthete: his hand-tooled injection kit marks him as a connoisseur as much as his meerschaum pipe and his Stradivarius violin. Doyle was at this time immersed in the 'yellow', decadent writings of Oscar Wilde, whom he met in 1889 at a dinner at the Langham Hotel in Bloomsbury hosted by J.M. Stoddart, editor of *Lippincott's Magazine*, which led to the publication of both *The Sign of Four* and *The Picture of Dorian Gray*. Holmes is a lover of art and pleasure, tantalisingly immune to public opinion; it seems likely that Doyle had Wilde, among others, in mind as he elaborated the bohemian character and lifestyle of his languid, cosmopolitan sophisticate. Holmes remains unusual among detectives in acting not out of a thirst for justice or compassion for the victims of crime, but simply a desire to keep himself amused. His detection is an art practised for art's sake.

As the Sherlock Holmes stories grew in popularity through the 1890s, the character's cocaine use came to suggest a seedy and disreputable edge that Doyle had not intended. In the early 1900s the Holmes stories were acquired by *Collier's*, a major player in the lucrative US magazine market selling over 250,000 copies a week. But *Collier's* was concurrently running a series of investigative pieces entitled 'The Great American Fraud' that exposed the prevalence of narcotics in patent medicines, a campaign that contributed to the regulation of the pharmacy trade by the 1906 Pure Food and Drug Act. A cocaine-injecting detective was out of step with the new times, and Doyle, who had in any case been mentioning his hero's drug use less and less, took steps to terminate it. In the short story 'The Adventures of the Missing Three-Quarter', published in *Collier's* in November 1904, Dr Watson announced that he had successfully 'weaned' Holmes from the 'drug mania' that had threatened to check his remarkable career.[86]

Further adjustments were made when the American actor-writer William Gillette brought Sherlock Holmes to the stage, first in the United States in 1899 and then a triumphant residency in London's Lyceum Theatre in 1901. The injection scene from the opening of *The Sign of Four* was included, but the exchange between Holmes and Watson was rewritten:

> WATSON: These drugs are poisons – slow but certain. They involve tissue changes of a most serious character.
> HOLMES: Just what I want! I'm bored to death with my present tissues and am out after a brand new lot!
> WATSON: Ah, Holmes – I'm trying to save you! (*Puts hand on HOLMES' shoulder*)
> HOLMES: (*Earnest an instant; places right hand on WATSON's arm*) You can't do it, old fellow – so don't waste your time.[87]

The languorous satisfaction of Holmes's habit is gone, replaced with the fatalistic self-pity of the doomed drug addict.

* * *

8. *The actor William Gillette in the 1899 Broadway production of*
Sherlock Holmes. *Gillette injects himself with cocaine while Dr Watson
(Bruce McRae) looks on disapprovingly.*

Cocaine was removed from Coca-Cola in 1905, and banned from general sale in the USA under the Harrison Narcotics Act of 1914 and in Britain by the Dangerous Drugs Act of 1920. In later life, Sigmund Freud rarely mentioned the 'cocaine episode', the term adopted by his biographer Ernest Jones. Jones's biography dispatched it with a brief, partial and evasive account; his phrase positioned it as a diversion from the main narrative of Freud's life and work, or perhaps a nervous breakdown to be passed over discreetly.

In *The Interpretation of Dreams* (1900) Freud wrote up his dream of 'Irma's Injection', in which he was confronted with a past medical error; the dream led him to recall the death of Fleischl, but he interpreted it to mean that he should allow himself to move on from his lingering guilt over the tragedy. The damning details of the Ernst Fleischl case were only fully exposed decades later, when copies of his long-classified letters to Martha were discovered in his archives.[88] In his *Autobiographical Study* of 1935 he accounted for his study of cocaine in one paragraph, referring to it as 'a side interest, though a deep one'. He suggested that Carl Köller had beaten him to the discovery of its anaesthetic properties because Freud had broken from his work schedule to visit Martha. As he recalled, he had been quite happy for Köller to have received the credit, 'and I bore my fiancée no grudge for the interruption of my work'.[89]

In these later writings, Freud made no reference to cocaine's stimulant and euphoric qualities, nor to any early insights the drug might have given him into the causal connection between mental states and physiological responses. Yet he never entirely lost his interest in the search for methods to increase the sum of nervous energy in his system. In 1923 he submitted himself to another, lesser-known experiment along these lines: the rejuvenation therapy developed by the Austrian endocrinologist Eugen Steinach, who claimed that his operation could restore its subjects to a 'second youth' by 'revitalising the glandular system'.[90] According to Steinach's theory, fatigue and loss of vitality in later life were caused by a decrease in testosterone; this could be arrested and reversed by a partial vasectomy in men, and by the application of X-rays to the ovaries in women. Just as the young Freud had theorised that cocaine might increase the reserves of energy in mind and body, Steinach argued that his 'vasoligation' procedure restored the potency of the seminal canals and raised the levels of vigour and energy available to the metabolic system.

Freud had been diagnosed that year with the cancer that would eventually claim him, and he may have been persuaded that the treatment would slow its progress.[91] Orthodox medical opinion regarded 'Steinaching' as a crank practice, along with the 'monkey-gland' therapies touted for rejuvenation at the time; in 1923 Conan Doyle published the classic Sherlock Holmes spine-chiller 'The Creeping Man', in which an ageing professor's experimental treatments with 'serum of anthropoid' cause him to scamper around the exterior walls of his house at night like a monkey.[92] Steinach's work was a continuation of the hormone therapy first trialled by Charles Brown-Séquard in his self-experiments with blood, semen and crushed testicles in the 1880s, while Freud had been investigating cocaine. As so often with pioneering self-experiments, it had taken a while for the productive applications of these researches to emerge. In 1894 'gland medicine' yielded its first promising drug, extracted from the adrenal gland and named adrenaline. It acted on the nervous system in a similar way to cocaine, dilating the pupils and boosting the action of the heart. Adrenaline spurred the search for new stimulant drugs, and in 1929 a synthetic compound with similar properties was discovered: amphetamine. With cocaine's medical use by this time mostly confined to local anaesthesia, amphetamine took on many of the applications that had been claimed for it in the 1880s: from battlefield stimulant to cognitive enhancer, diet pill to anti-depressant. Before long its medical promise was undermined by its non-medical use as a drug of pleasure, much as cocaine's had been half a century before.

In 1929, the same year that amphetamine was synthesised, Freud reflected in his *Civilization and Its Discontents* on whether such drugs were remedies for the disease of civilisation or symptoms of it. 'What decides the purpose of life', he wrote, 'is simply the programme of the pleasure principle'; yet the pursuit of pleasure cannot avoid being forced into conflict with reality.[93] In this age-old struggle, 'the crudest, but also among the most effective among the methods of influence is the chemical one – intoxication'. Drugs give us an edge over our circumstances: 'not merely the immediate yield of pleasure, but also a greatly desired degree of independence from the external world', by offering temporary escape from pain, fatigue or boredom.[94] Yet they can never actually resolve our conflicts with the external world, and they often end up weakening our ability to manage them. In themselves they are neither disease nor cure; their benefits and dangers are, as James Lee maintained, a function of how we deploy them. They have played an enduring and intimate role in civilisation's control over the forces of nature: they are part of a formidable toolkit that extends from the mastery of fire to the modern technologies that extend our mental reach through everything from writing to photography to the telephone. In the modern age:

> Man has, as it were, become a kind of prosthetic god. When he puts on all his auxiliary organs he is truly magnificent; but those organs have not grown on to him and they still give him much trouble at times . . . Future ages will bring with them new and probably unimaginably great advances in this field of civilisation and will increase man's likeness to God still more. But in the interests of our investigations, we will not forget that present-day man does not feel happy in his God-like character.[95]

As Freud came to recognise through his psychoanalytic practice, one of the most intractable obstacles to therapy was the wish of his patients not to be cured. Pain and pleasure were not so easily separated; in their different ways, both gratifying our instincts and relinquishing our suffering are difficult and frightening prospects.

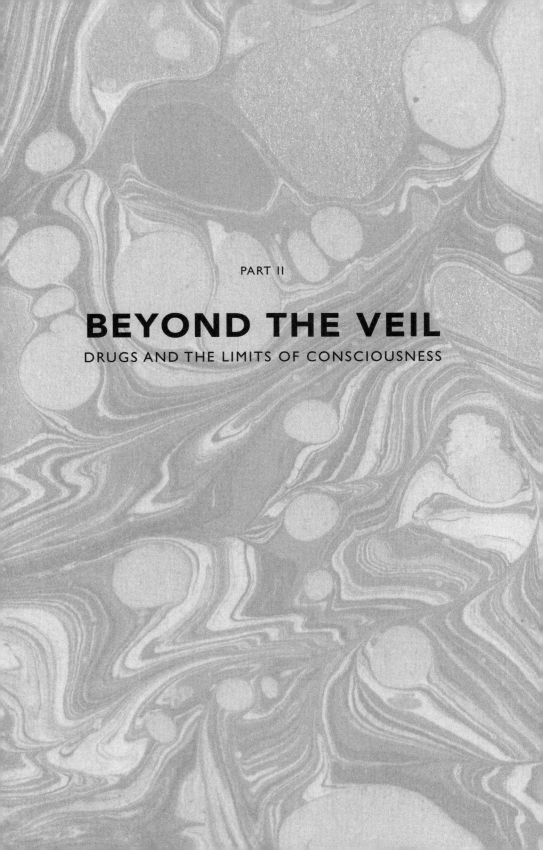

PART II

BEYOND THE VEIL

DRUGS AND THE LIMITS OF CONSCIOUSNESS

9. 'Ether Dreams', illustration for Les Merveilles de la science, *a popular work by the chemistry professor Louis Figuier (1868).*

CHAPTER THREE

A WORLD OF PURE EXPERIENCE

In 1882 the forty-year-old William James, recently appointed assistant professor of philosophy at Harvard, set up a private experiment in the university's chemistry laboratories. Following the procedure described by Humphry Davy at the Pneumatic Institution nearly a century earlier, he gently heated a glass beaker of ammonium nitrate, captured the escaping nitrous oxide in a gas holder and inhaled it with a pen and notebook in front of him. A few minutes – or perhaps an eternity – later, he found himself staring at 'sheet after sheet of phrases dictated or written during the intoxication'.[1]

The experiment imprinted itself indelibly on James's mind, and remained vivid for decades. What it meant, though, was a question with which he would struggle for the rest of his life. 'The keynote of the experience,' he wrote in his first published description, 'is the tremendously exciting sense of an intense metaphysical illumination. Truth lies open to the view in depth beneath depth of almost blinding evidence.'[2] He included a long string of the jottings he made under the influence in an addendum to his short paper on G.W.F. Hegel's philosophy in *Mind*, one of the leading journals of the new psychology:

Reconciliation of opposites; sober, drunk, all the same!
Good and evil reconciled in a laugh!
It escapes, it escapes!
But —
What escapes, WHAT escapes?

. . .

There *is* a reconciliation!
Reconciliation — econciliation!
By God, how that hurts! By God, how it *doesn't* hurt!
Reconciliation of two extremes.
By George, nothing but othing!
That sounds like nonsense, but it is pure *on*sense![3]

'To the sober reader', he conceded, this may 'seem meaningless drivel', and yet he made no apology for submitting it to a learned journal. 'At the moment of transcribing', he recalled, it was 'fused in the fire of infinite rationality'. If it struck the reader as absurd, it was because they had never experienced this reconciliation of opposites for themselves.[4] 'I strongly urge others to repeat the experiment,' he added, 'which with pure gas is short and harmless enough.'[5]

James had been persuaded to self-experiment by a privately printed pamphlet, *The Anaesthetic Revelation and the Gist of Philosophy*, published in 1874 by Benjamin Blood. Blood was a native of Amsterdam in upstate New York, with no scholarly credentials but a colourful portfolio of occupations that included engineer, patent developer, gambler, bodybuilder, prize-fighter, calculating prodigy and energetic contributor to the letters pages of the local newspapers. In the short review he had written for the *Atlantic Monthly* on its publication, James professed himself initially 'more than sceptical of the importance of Mr Blood's so-called discovery' and predicted that 'crack-brained, will be the verdict of most readers'.[6] Yet he had been captivated by the majestic confidence of Blood's oratory – 'You have the greatest gift of superior gab since Shakespeare,' he told him later[7] – and resolved that he would not 'howl with the wolves or join the multitude in jeering at it'.[8] Later he would recognise that it had been a crucial stepping stone to his mature philosophy, and it remained 'one of the cornerstones or landmarks' of his view of the mind up to and including his final published work.[9]

10. The young William James in Brazil, 1865.

Blood first encountered nitrous oxide in 1860 in the same way that many thousands of Americans did over the course of the nineteenth century: 'through the necessary use of anaesthetics' in the dentist's chair. When he returned to consciousness, reeling from his entirely unexpected 'Revelation or insight of the immemorial Mystery', he quizzed doctors and dentists for an explanation of his epiphany:[10]

> I learned that nearly every hospital and dental office has its reminiscences of patients who, after a brief anaesthesia, uttered confused fragments of some inarticulate import which always had to do with the mystery of life, of fate, continuance, necessity and cognate abstractions, and all demanding, 'What is it? What does it all mean, or amount to?'[11]

Blood received little answer beyond smiles and shrugs, but he was unable to set these momentous questions aside. He sought the gas out for further experiments at regular intervals, until after fourteen years he had elaborated his anaesthetic revelation into a full-blown philosophy. He persuaded James to follow his method with the same argument by which James sought to persuade his own readers: that this insight was of a different order from the usual categories of philosophy and could only be gained by direct experience. 'By the Anaesthetic Revelation', Blood wrote:

> I mean a certain survived condition (or uncondition,) in which is the satisfaction of philosophy by an appreciating of the genius of being, which appreciation cannot be brought out of that condition into the normal sanity of sense – cannot be formally remembered, but remains informal, forgotten until we return to it.[12]

His revelation, he insisted, was not obscure or technical; it was merely hard to bring it back from the place in which it was to be found. 'The plain truth,' he wrote, 'is that the modern student of philosophy has been baffled, daunted and discomfited by a fake esotericism, arbitrarily technical in terms and presumptions.' In reality the secret of the universe was wide open to us, yet only in certain states of consciousness – a situation Blood illustrated by quoting from Tennyson:

As here we find in trances, men
Forget the dream that happens then,
Until they fall in trance again[13]

The truth was revealed not while passing from the waking state into the euphoric, dissociated embrace of the gas, but at the moment when waking consciousness was returning. Those awakening from it in the dentist's chair often glimpsed it as it diffused and dispersed, leaving only a faint and ebbing sensation of having visited another world. Blood, through his self-experiments, had learned to make sense of it by applying the precepts of German idealist thought, particularly that of G.W.F. Hegel.

'The inveterate or abiding knot or twist which baffles philosophy', he explained, was the question of 'identity and difference'.[14] The world appears to us as a single unified whole, but as soon as we begin to describe it we set its component parts in opposition to each other: mind and matter, reality and appearance, reason and emotion, something and nothing. Reality as we perceive and describe it is created out of these oppositions, and philosophers from the ancient Greeks onwards have viewed the conflict between them as the motive force of the universe. Hegel, however, showed that opposites can be seen not as the building blocks of reality but as a stage in a process, thesis and antithesis: at a higher level of observation, they resolve into a unity, presided over by the spirit of the Absolute that unites mind and matter. This dissolution of opposites cannot be achieved by reason, since it transcends the principles of binary logic. It can, however, be experienced directly under the influence of nitrous oxide, where opposites coexist without contradiction: 'The One remains, the many change and pass.' It is a philosophy, but equally an ecstatic state presided over by the radiant light of Christ, and Blood conveyed it with the oracular force of prophecy:

> This world is no more that alien terror which was taught me. Spurning the cloud-grimed and still sultry battlements whence so lately Jehovan thunders boomed, my gray gull lifts her wing against the nightfall, and takes the dim leagues with a fearless eye.[15]

* * *

Humphry Davy, on Boxing Day of 1799, had been the first to discover that inhaling nitrous oxide could create a state of consciousness in which the mind appeared to exist independently of the material world. He had taken his self-experiment to the limits by sitting in an enclosed box, designed by the engineer James Watt for the use of patients too weak to inhale, and having it filled with 20 quarts of gas every 5 minutes. After an hour and a quarter, when he felt he had saturated his system, he stepped from the box and inhaled a further 20 quarts of gas from a giant silk air-bag. At this point, he recorded that he 'lost touch with all external things' and was projected into a world composed entirely of mental events. He 'existed in a world of newly connected and newly modified ideas' that combined 'in such a manner, as to produce perceptions perfectly novel'.[16]

Until this point, such states had been experienced only *in extremis*: in visions, possessions or ecstasies, during physical trauma or seizures, as the result of rigorous and extensive spiritual practice, or at the point of death. Now, they could be experienced at will by anyone with a basic chemical apparatus. It was curious that this state of consciousness should become available experimentally just as German idealist philosophy was making its passage into the Anglophone world. Thomas Beddoes, Davy's employer, was a voracious German reader and in 1793 had written perhaps the first British commentary on Immanuel Kant;[17] in 1799, Samuel Taylor Coleridge had just returned from studying in Germany when he and Davy forged their deep friendship over a green silk bag of the gas. Kant's philosophy proposed that human experience and knowledge was embodied, bounded by the senses and hardwired categories of thought such as time and space. Under nitrous oxide, however, as the physical world receded, these categories were dissolved and the mind freed to explore what William James would later call 'a world of pure experience'.[18]

Davy's revelation welled up in him, just as Blood described, at the moment when he was returning to his normal waking state. The words he spoke 'in the most intense and prophetic manner' echoed the central tenet of the new German philosophy: 'Nothing exists but thoughts! The universe is composed of impressions, ideas, pleasure and pains!'[19] Like many famous eureka moments in science, his prophetic utterance may have been less spontaneous than he claimed, artfully composed at a later moment. Nonetheless, it became the best-remembered and most quoted drug epiphany of the nineteenth century. It was rarely noted, though, that the two sentences expressed quite different philosophies. The second stated a theory long familiar to Davy, Beddoes and their circle and often referred to as 'sensationism' or 'associationism': that ideas arose from sensory stimuli, as clusters of thought and feeling formed by repetition and habit. But the first, that 'nothing exists but thoughts', was a far more extreme and striking claim. In Davy's account, it suggests the influence of Coleridge, who was at that time in thrall to the idealist philosophy of Bishop Berkeley, and it formed the ground on which Benjamin Blood raised his theory seventy years later.

In the interim, experiments with nitrous oxide had followed a circuitous route to the drug's defining application in the dentist's chair. Initially, many attempted to follow Davy to his world of pure thought, but few succeeded. His epiphany had been underpinned by the Pneumatic Institution's visionary project, with its effusive sense of possibility; those who attempted the experiment outside Bristol, even

Beddoes and Davy's close allies such as James Watt, were 'much less affected' than they had hoped. Watt was left wondering if he had synthesised the wrong gas by mistake.[20] Even among Day's volunteers, many had a quite different experience. The novelist Maria Edgeworth, a wry observer at the Pneumatic Institution's experiments, noted that attempts by volunteers often produced 'nothing but a sick stomach and a giddy head'. Despite Davy's pioneering use of placebo in the form of a bag of normal air, she reckoned that expectation played a large part in the experience. 'Faith, great faith,' Edgeworth concluded, 'is I believe necessary to produce any effect.'[21]

Those who managed to attain the promised exhilaration were, as most of Davy's volunteers had been, more struck by the unbidden rush of pleasure than any metaphysical transcendence. In most cases their descriptions echoed that published by Davy, and the consensus view of the experience that emerged drew heavily on his most famous phrases. The first recorded experiment in the United States in 1808, by the medical student William Barton – future professor of botany at the University of Pennsylvania – merged Davy's insight with Robert Southey's encomium:

> My lungs felt as if they were dilating, and they continued to impart this sensation of enlargement till I supposed they occupied the whole laboratory with their immensity. I now became totally insensible to the impressions of external things, and the rapturous delight which then entranced my faculties, mars my feeble essay towards its description. This indescribable extacy [sic] must be what angels feel; and well might the poetick Southey exclaim upon experiencing it, that 'the atmosphere of highest of all possible heavens must be composed of this gas'.[22]

Despite such spectacular reports, nitrous oxide was unable to establish a compelling medical application. Davy anticipated its eventual use in surgical anaesthesia just as Freud did with cocaine: he included in his avenues for future research the proposition that 'as nitrous oxide in its extensive application appears capable of destroying physical pain, it may probably be used with great advantage during surgical operations'.[23] But this suggestion went unheeded for half a century. The majority of surgeons remained firmly opposed to anaesthesia until the 1840s, in part for practical reasons – the operating theatre was no place for flammable chemical reactions – and in part due to the prevailing medical theory that pain could not be eliminated from surgery, since keeping patients conscious was essential for keeping them alive.

In the meantime, scientific experiments with the gas devolved into the pursuit of its curious pleasures. The young German naturalist Christian Friedrich Schönbein, who would go on to discover ozone and invent the fuel cell, spent some months of 1826 in Britain where he stayed with a chemical experimentalist and made 'a good supply' of nitrous oxide in his home laboratory. One sunny afternoon, a large party met in the garden to inhale the intoxicating gas in the open air. Several young men did so with 'undoubted signs of wellbeing and pleasure', prompting a more mature gentleman, in his turn, 'to dance and devastate the adjoining flower-bed in his ecstasy – to the delight of his audience'. Schönbein was struck by the peculiarly British combination of high-flown philosophising and ribald humour, and speculated: 'Maybe it will become the custom for us to inhale laughing gas at the end of a dinner party, instead of drinking champagne.'[24]

The enduring nickname 'laughing gas' was in use by 1824, when it featured on the poster for a variety evening at London's Adelphi Theatre offering 'Uncommon Illusions, Wonderful Metamorphoses, Experimental Chemistry, Animated Paintings etc.'[25] The conventions of the public nitrous oxide demonstration were rapidly established. A master of ceremonies, usually adopting the persona of a scientific educator, would present the laboratory equipment, describe the chemical properties of the gas and ask for volunteers to inhale it. Often the first pick would be a collaborator who would scoff loudly before submitting himself (women were not permitted) to a lungful of gas

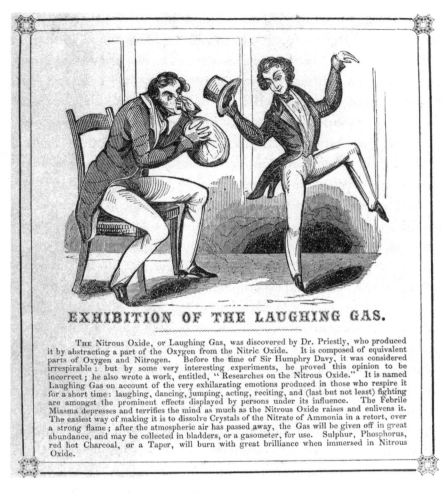

EXHIBITION OF THE LAUGHING GAS.

THE Nitrous Oxide, or Laughing Gas, was discovered by Dr. Priestly, who produced it by abstracting a part of the Oxygen from the Nitric Oxide. It is composed of equivalent parts of Oxygen and Nitrogen. Before the time of Sir Humphry Davy, it was considered irrespirable : but by some very interesting experiments, he proved this opinion to be incorrect ; he also wrote a work, entitled, " Researches on the Nitrous Oxide." It is named Laughing Gas on account of the very exhilarating emotions produced in those who respire it for a short time : laughing, dancing, jumping, acting, reciting, and (last but not least) fighting are amongst the prominent effects displayed by persons under its influence. The Febrile Miasma depresses and terrifies the mind as much as the Nitrous Oxide raises and enlivens it. The easiest way of making it is to dissolve Crystals of the Nitrate of Ammonia in a retort, over a strong flame ; after the atmospheric air has passed away, the Gas will be given off in great abundance, and may be collected in bladders, or a gasometer, for use. Sulphur, Phosphorus, red hot Charcoal, or a Taper, will burn with great brilliance when immersed in Nitrous Oxide.

11. *Nitrous oxide acquired its enduring nickname 'laughing gas' in the 1820s, when self-experiments with it became a staple of popular entertainment.*

and proceeding to behave outrageously – pirouetting, spouting poetry, spoiling for a fight – to the delight of the audience. More volunteers would follow, each attempting to outperform the last.

The effects of the gas proved far more reliably spectacular in this context than in the laboratory. The first recorded nitrous show in the

United States, in Philadelphia in 1814, was described by a local printer and pamphleteer, Moses Thomas, and gives a vivid sense of the proceedings. The experiment was part of a weekly lecture series, and the stage that evening was blocked off from both the laboratory equipment and the audience, heightening expectations of the unexpected. It opened with a doctor discoursing on 'the nature and properties of nitrous oxide' and demonstrating 'a number of unimportant experiments, to which very little attention is paid by his auditors' before he launched into the promised spectacle. The first volunteer, a young man of fifteen, was presented with 'a large bladder' filled with 'the exhilarating gas', which he inhaled, and then he

> suddenly threw away the bag, with an air of triumphant disdain, and began to march about the inclosure with theatric strides, until coming up close to the front row, he perceived that one of the persons sat there held a cane athwart to defend himself from his too near approach. This offended his pride – he instantly burst into a paroxysm of rage: 'That tyrant', says he, 'has seized my cane – deliver it to me – this – instant – or I'll be the death of you!' At the same moment jumping over the desk, and grappling with the man who had the cane, he overturned every thing that stood in his way, and it required the united efforts of four or five men to hold him down, till the effect of the gas ceased, and he turned round to the company with an air of good-humoured hilarity.[26]

More volunteers followed, and exhibited 'different degrees of animation, or ferocity, dancing, kicking, jumping, fencing, and occasionally boxing anyone that stood in their way'.[27] This kind of 'antic' behaviour had been a regular feature of the Pneumatic Institution's experiments: volunteers had often leapt about, rushed up and down stairs or launched into a poem or diatribe, before returning to consciousness with no recollection of their bizarre speech or actions. The gas allowed the conscious mind to escape from the shackles of the body, but in so doing it left behind a body at the whims of unpredictable or automatic forces. These often seemed to express a second and hidden personality, which the popular lecture format mined for entertainment in a manner similar to the stage hypnotism shows of today. On stage, the subjective experience was incidental; the setting was constructed to make the most of its external manifestations. The moment of return to waking to consciousness was not interrogated for mystical revelation, but held up for confused hilarity.

Nitrous oxide was ideally suited to the travelling carnival-show circuit that expanded across the wide and sparsely populated interior of ante-bellum North America and Canada. It could be presented as both a miracle of modern science and a raucous entertainment, a novelty, both mind-boggling and edifying, that could play out every night for years in front of a fresh and astonished crowd. The gas tank, tubes and breathing bags required a capital outlay that placed the entertainment beyond the reach of private individuals, but thereafter the profits per dose of gas were considerable. Robert Southey's line about 'the atmosphere of heaven', the perfect advertising slogan, was frequently blazoned across marquees and posters. Entrepreneurs, innovators, medicine men, hucksters, grifters, mountebanks and frontier individualists of all kinds turned their hands to it. In 1832 the eighteen-year-old Samuel Colt toured a nitrous oxide show from Canada to Maryland, complete with mobile laboratory, under his old family name of 'Coult': he dosed an estimated 20,000 volunteers at 25 cents a time, using his takings to fund trials and prototypes of his repeating revolver pistol. 'The sensations produced by it are highly pleasurable', Colt wrote in his advertisement for a show at Portland, in which he announced 'select' demonstrations for ladies, at which 'not a shadow of impropriety attends the exhibition'.[28]

*　　*　　*

It was in this seamy but energetic milieu of inventors and dream sellers that nitrous oxide found the medical application that had eluded the philosophers of the Pneumatic Institution. The figure who first turned it from an agent of pleasure to a conqueror of pain was a flamboyant medical lecturer named Gardner Quincy Colton, whose original show-piece was a spectacle named 'Court of Death', a giant diorama canvas depicting the evils of drink and the salvific virtues of temperance. He set the scene up nightly in revival tents, Masonic halls, hotels and lecture theatres to illustrate his lecture on the subject, charging the standard fee of 25 cents a ticket. In the early 1840s he added nitrous oxide to his repertoire, introducing it with a lecture that reinforced his theme. The gas, he claimed, offered a living demonstration of original sin, and would reveal the true character of anyone who inhaled it.

Colton adopted the usual theatrical precautions – barriers around the stage, strongmen posted in the front row, 'safe' dress seats for the ladies – and assured the crowd that the proceedings would be conducted 'with the propriety and decorum which shall deserve the patronage of an intelligent class of ladies and gentlemen'.[29] Suitably primed, the volunteers from the audience acted out what were supposed to be their deepest instincts and desires: fighting, singing, laughing, dancing, declaiming, giving orders to the audience or blurting out secrets. Colton would often guess in advance how each volunteer would react, an effec-tive combination of suggestion and challenge. The method was, he claimed, a more accurate guide to personality even than phrenology.

It was during an evening performance in 1844 in Hartford, Connecticut that Colton's show became the improbable inspiration for the great medical discovery of the age. It was announced in a local newspaper with Robert Southey's famous quotation, and the assurance that 'eight strong men are engaged to occupy the front seats, to protect those under the influence of the gas from injuring themselves or others'.[30] Nonetheless, one of the audience volunteers, a young drug-store clerk, thrashed around wildly under the influence of the gas and crashed into a bench; he suffered a severe gash and bleeding but, as a local dentist in the audience noticed, no accompanying pain. Horace Wells, the dentist, inspected Colton's equipment after the show: a simple rubber bag attached with a wooden faucet to something resembling a cider barrel filled with nitrous oxide. Like all dentists, Wells's business was limited by the amount of pain his customers were prepared to endure, and at this moment he was himself suffering from an excruciating impacted wisdom tooth. He asked Colton to administer the gas to him while another dentist removed it. Wells returned to consciousness with no sensation of pain, and a revelation of more practical value than an encounter with the Hegelian Absolute. His prophetic utterance was: 'A new era in tooth-pulling!'[31]

Wells's self-experiment set in motion the discovery of what was shortly to be christened anaesthesia. It was an era-defining triumph for modern medicine, but it emerged from a series of unedifying jealousies, intrigues, squabbles and tragedies. In 1845 Wells persuaded Boston's leading surgeon, John Collis Warren of Massachusetts General Hospital, to allow him to administer the gas during an operation. All was proceeding to plan until the patient groaned loudly during the procedure; Wells was booed from the theatre, though it turned out afterwards that the patient had been fully unconscious and the sound was involuntary. Wells never recovered from the humiliation, lapsing into a depression and eventually taking his own life. Meanwhile his associates and rivals in the small world of Boston dentistry lost no time in making their own experiments. His former partner, William Thomas Green Morton, consulted a local chemist named Charles Jackson about possible alternatives to nitrous oxide, and alighted on diethyl ether.

Ether is a volatile compound, halfway between gas and liquid at room temperature, that had been familiar to chemists for centuries as a solvent. The German physician Valerius Cordus described its synthesis from ethyl alcohol and sulphuric acid in 1540, and gave it the name 'sweet oil of vitriol'. 'It is so sweet', the physician and alchemist Paracelsus wrote around the same time, 'that chickens eat it and then fall asleep, but wake up after some time without any bad effect'. Paracelsus also noted that 'it extinguishes pain', and it had been commonly used in medicine since the eighteenth century:[32] its cool evaporation relieved lung conditions when it was inhaled, and some doctors also recommended it as a nervous tranquilliser. Morton began his experiments on animals, and observed that a dog 'wilted away' when its head was held over an ether bowl.[33] He proceeded to inhale it himself from a soaked handkerchief, and worked his way up to a dose that kept him unconscious for seven or eight minutes. In September 1846 he experimented with it on one of his customers, and removed a tooth painlessly.

The following month Morton followed Horace Wells into the operating theatre of Massachusetts General. He had persuaded John Collis Warren that his new compound was far superior to nitrous oxide, and in an attempt to patent his discovery he presented it as a novel substance he called 'letheon'. It was to be administered via a proprietary device he named 'the cone of oblivion': essentially a glass funnel with an ether-soaked sponge in the bottom, to be held over the patient's face. Morton's sales pitch was met with scepticism from the assembled physicians, but he succeeded where Wells had failed. Warren's patient, a twenty-year-old man with a congenital malformation in his neck, inhaled from the cone of oblivion and sank into a deep slumber. Morton turned to Warren with the words 'Your patient is ready, doctor'; as he made the first incision, Warren raised his head to address the crowded tiers of benches: 'Gentlemen, this is no humbug.' After the procedure, the patient testified: 'I did not experience pain at any time, though I knew that the operation was proceeding.'[34] It was a famous moment, commemorated by renaming the theatre 'The Ether Dome', a name it retains proudly to this day. Every 22 October it becomes the focus for Ether Day, where commemorative ribbons and pins are issued and long-serving hospital staff honoured.

The outcome for the protagonists was less triumphant. Morton persevered for years with patent applications, which were fiercely contested by Charles Jackson, who claimed the idea of using ether had been stolen from him. Jackson applied to the French Academy of Sciences for the reward of 5,000 francs offered for great discoveries, which he was granted on condition he shared it with Morton. Morton

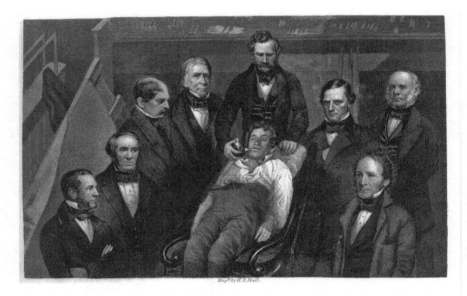

12. The celebrated operation at Massachusetts General Hospital on 22 October 1846 which demonstrated the efficacy of ether as a surgical anaesthetic.

refused, claiming the status of sole inventor, and eventually persuaded Congress to issue an Appropriation Bill for $100,000 in his favour. This was contested in turn by Horace Wells's widow, who claimed that her late husband was owed a share in the discovery, and by supporters of a Georgia surgeon named Crawford Williamson Long who, it emerged, had been using ether on his patients for several years previously. The Senate overturned Morton's award on grounds of multiplicity of claimants, after which Morton bankrupted himself with fruitless legal suits against Jackson, who was himself embroiled in legal wrangles with another of his former associates, Samuel Morse, over the invention of the telegraph. Meanwhile, Gardner Quincy Colton abandoned the 'Court of Death' and reinvented himself as the Colton Dental Association, setting up a string of offices and surgeries along the east coast. By 1886 he claimed to have performed 125,000 painless tooth extractions under nitrous oxide.

*   *   *

William Morton's opportunistic attempt to rebrand ether as 'letheon' failed as soon as its sharp solvent reek reached John Collis Warren's nostrils. Warren recognised it immediately, not only from the surgeon's dispensary but from the 'ether frolics' among medical students that, as he recalled, 'often figured conspicuously at the close of a course of medical lectures in certain parts of the country'.[35] These had grown out of the vogue for nitrous oxide parties in the early years of the nineteenth century, and taken hold because ether was simpler to manage: rather than chemical reactions, tubes, pumps and inhalers, it required only a bottle of ether and a cloth to soak and hold under the nose.

'Ether frolics' combined experiment and recreation, with a character somewhere between the expansive antics inspired by nitrous oxide and the boisterous camaraderie of strong alcohol shots. A few deep inhalations produced a sense of swelling in the head, ringing in the ears, a numb palate and a rushing crescendo of intense confusion that, like nitrous oxide, could resolve into a revelation of startling clarity or an uncanny sense of *déjà vu*. Ether is more toxic than nitrous oxide, and an extended period of inhalation leads to stumbling, collapse and blackouts, from which the subject recovers swiftly but usually with a hangover, nausea, raw skin around the nose and a pounding headache. It was after a session of this kind in Athens, Georgia that the surgeon Crawford Long, who with 'several of my young associates frequently assembled ourselves together and took it for the excitement it produces', noticed that he was covered in bruises that he had no recollection of incurring and began – allegedly in 1842 – to experiment with it in surgical operations.[36]

Despite its momentous contribution to surgery, the disadvantages of ether were apparent from the start. It was highly flammable, it stank like petrol, it produced irritant rashes and the side-effects of an extended inhalation on a fragile patient were brutal. The search for an alternative promised great rewards. One of ether's early champions in Britain was James

Young Simpson, professor of medicine and midwifery at the University of Edinburgh. British surgeons were initially sceptical of the miracle drug, and indeed of the very idea of a medical breakthrough emerging from the United States: Robert Liston, London's most distinguished surgeon, referred to it as a 'Yankee dodge'.[37] But its miraculous results were plain to see. Liston maintained that it was only useful for short operations; Simpson, who saw that it could make more complicated procedures possible, worked with it intensively in his Edinburgh surgery. During the summer and autumn of 1847, he enlisted the services of chemists and took on an assistant, a younger physician named James Matthews Duncan. The morning room of Simpson's home was soon filled with pharmacy bottles, and the pair spent evenings tipping their contents into saucers, inhaling them and making notes on their effects.

Simpson and a revolving cast of assistants worked their way through a variety of solvents including acetone (which proved toxic to the lungs), ethyl nitrate (splitting headache), benzene (ringing in the ears), carbon disulphide (headaches and revolting smell), before a chemist of his acquaintance suggested chloroform, or 'chloric ether', which had previously been used in the manufacture of gunpowder mixes. On 4 November 1847 Simpson and his colleagues, together with members of his family, sat round the dining table and inhaled the vapour slowly from tumblers. The smell was sweet, with the cool burn of liquorice or menthol, and as they inhaled their eyes sparkled and animated conversation flowed. 'This is far better and stronger than ether', Simpson recalled thinking – and then 'we were all under the mahogany in a trice'.[38] He woke to find himself on the floor, staring up at the underside of the table and listening to a series of heavy thumps as his companions collapsed beside him on the carpet. They were so delighted they decided to try it again.

Because they took place in domestic surroundings, Simpson's self-experiments are unusual in including the testimony of women. His wife, her sister and her daughter were all present for the famous experiment. His wife's niece, Agnes Petrie, is recorded as sitting at the table,

13. *James Young Simpson, surgeon and professor of midwifery at the University of Edinburgh, discovered chloroform's anaesthetic properties by self-experimenting in his living room.*

inhaling the chloroform, folding her arms and falling asleep while moaning 'I'm an angel! Oh I'm an angel!', and on waking, 'How are you all down there?'[39] They participated in several sessions over the summer, on one occasion witnessed by a visiting celebrity, Hans Christian Andersen. Andersen visited Scotland in August 1847 and Simpson extended a dinner invitation to 'the Danish Walter Scott', as the local newspapers referred to him. After dinner, to Andersen's surprise, ether was passed round:

> I thought it distasteful, especially to see ladies in this dreamy intox- ication; they laughed with open, lifeless eyes; there was something unpleasant about it, and I said so, recognising at the same time that it was a wonderful and blessed invention to use in painful opera- tions, but not to play with it; it was wrong to do it; it was almost like tempting God.[40]

The mastery of anaesthetics over pain cast a fresh light on their use for pleasure. Initially, the two faces of the drugs were seen as complementary. John Collis Warren thought it a delightful paradox that the surgical torture of the body was now 'accompanied by a delightful dream' or a 'celestial vision'. Other physicians were struck by the philosophical questions that had fascinated Davy: 'What extraordinary phenomena it presents! The understanding awake and conscious – the body impassive. The soul becoming almost a stranger to the body, even in this life.'[41] As anaesthesia spread rapidly around the globe, such mysteries were widely discussed. 'The personal experiences of the etherised are most interesting', an 1847 report in the Scottish periodical the *People's Journal* noted; 'The mind appears perplexed what to make of its new condition', floating free of the body with the 'feeling as if the chains which bind it to the earthly tenement were burst'. Many subjects experienced sensations 'of the most curious and thrilling character'. For some it was an ecstatic moment, a feeling that their soul was bathed 'in an atmosphere of light, and revels in the happiness of apparently another world', leaving an evanescent afterglow that faded on awakening. Others, however – and it was impossible to predict who they would be – had a quite different experience, finding themselves hurled into a waking nightmare:

> Some are violently altercating with their relations; some are tumbling down bottomless abysms; and horrors of all sorts crowd in on the minds of others . . . Where this state exists, the termination of the process is an inexpressible joy to the etherised, but it is an almost universal experience that no actual pain was felt: – 'It was like a frightful dream!', or 'I thought a fiend had hold of me!'[42]

Once the novelty of ether receded, however, subjective experiences of this kind were rarely recorded in the medical literature. Their causes and meanings remained a mystery, as did the reasons why the state was experienced so differently by different subjects. But clinical descriptions of visions became shorter and more formulaic, and the subjective experience was more commonly compared to drunkenness. John Collis Warren characterised it as a 'delirium', a side-effect that signified no more to the physician than the mental confusion of a high fever. He was soon warning that the pleasures of ether and chloroform were a dangerous invitation to abuse. In his 1849 pamphlet, *Effects of Chloroform and of Strong Chloric Ether, as Narcotic Agents*, he wrote: 'We were soon awakened from our dreams of the delightful influence of the new agent, by the occurrence of unfortunate and painful consequences.'[43]

One of these consequences was the discovery that, in rare cases, chloroform could stop the heart and render the delightful dream tragically permanent. More common, however, was the use of the new drugs as remedies for nervous excitement, and the recognition that this could become a chronic, escalating and destructive habit. 'Many persons of both sexes', Warren recorded, 'have resorted to it for the purpose of obtaining the pleasure of a temporary delirium.'[44] He cited the case of:

> a physician in New York, who had achieved some distinction, and who suffered from mental trouble, resort[ing] to the use of chloroform so frequently as ultimately to produce a delirium; during which he committed some extraordinary acts, and finally destroyed himself.[45]

Warren is referring to Horace Wells, who took to using chloroform habitually during the depression that followed his abortive attempt to demonstrate nitrous oxide in Warren's operating theatre. Wells had moved from Hartford to New York, where he wandered the streets in a confused state under its influence, and in 1848 was arrested for throwing sulphuric acid on the shawls of two sex workers in the street. Arrested and jailed, he took his own life by cutting open his femoral artery. He was perhaps the first casualty of the new vice, but Warren stressed that 'the number of individuals whom I have known to use it in this way is so considerable as to lead me to believe, that those who employ it in secrecy must be very great'.[46] As another doctor observed, pleasure was not simply a curious by-product of the new anaesthetic drugs, but a menace to public health: they 'delight the animal sensations, while they destroy the moral sentiments; they introduce their victims to a fool's paradise; they mock them with joys which end in sorrows'.[47]

The shift away from subjectivity was epitomised by Oliver Wendell Holmes – doctor, lawyer, poet and dedicated polymath, Harvard professor and intimate of William James's family – who in 1846, in correspondence with Henry Morton, coined the term 'anaesthesia'. Various terms had initially competed, but all were either too specific ('etherisation'), too broad ('analgesia') or too elaborate ('etherial narcotism'). Holmes felt that such a momentous discovery deserved an addition to the language. 'Anaesthesia', he suggested, 'signifies insensibility', as opposed to analgesia, which merely denoted the absence of pain. He also proposed that 'the adjective will be anaesthetic'.[48] In a lecture to the Phi Beta Kappa Society at Harvard in 1870, Holmes delivered a much-quoted anecdote:

> I once inhaled a pretty full dose of ether, with the determination to put on record, at the earliest moment of regaining consciousness, the thought I should find uppermost in my mind. The mighty music of the triumphal march into nothingness reverberated through my brain, and filled me with a sense of infinite possibilities, which made me an archangel for the moment. The veil of eternity was lifted. The one great truth which underlies all human experience, and is the key to all the mysteries that philosophy has sought in vain to solve, flashed upon me in a sudden revelation. Henceforth all was clear: a few words had lifted my intelligence to the level of the knowledge of the cherubim. As my natural condition returned, I remembered my resolution; and, staggering to my desk, I wrote, in ill-shaped, straggling characters, the all-embracing truth still glimmering in my consciousness. The words were these

(children may smile; the wise will ponder): 'A strong smell of turpentine prevails throughout.'[49]

Holmes's bathetic parody of the anaesthetic revelation is better remembered than most of the experiences it referenced, thanks to its repetition in prominent twentieth-century sources such as Bertrand Russell's *A History of Western Philosophy* (though Russell attributed the line to William James, substituted laughing gas for ether, and misquoted turpentine as petroleum).[50] It assures the reader in urbane terms that claims of visionary experience under anaesthetics are delusional: grandiose fantasies that wither to nothing under the cold light of reason. It was a message that reassured the rising generation of Harvard medical students, and resonated with Boston's venerable philosophical traditions. The Transcendentalists, despite their valorisation of direct experience, their love of nature and their curiosity about the spiritual insights of Buddhism, regarded artificial intoxicants as an untrustworthy and corrupting influence. Ralph Waldo Emerson, a friend of Oliver Wendell Holmes and of Henry James senior, William's father, made his presumption against 'procurers of animal exhilaration' explicit in his 1846 essay 'The Poet':

Never can any advantage be taken of nature by a trick. The spirit of the world, the great calm presence of the Creator, comes not forth to the sorceries of opium or of wine. The sublime vision comes to the pure and simple soul in a clean and chaste body. This is not an inspiration, which we owe to narcotics, but some counterfeit excitement and fury.[51]

Henry Thoreau, writing in his diary after a dental procedure in 1851, took a similar view. Ether, he wrote, offered no authentic 'spiritual knowings' to those 'who attend to the intimations of reason or conscience'. Nonetheless, the experience of disembodiment was a singular one:

By taking the ether the other day I was convinced how far asunder a man could be separated from his senses. You are told it will make you unconscious – But no-one can imagine what it is to be unconscious: how far removed from the state of consciousness and all that we call 'this world' – until he has experienced it . . . You are a sane mind without organs – groping for organs – which if it did not recover its old sense would get new ones.[52]

It was an uncanny experience, a disembodied brain searching for stimuli in the void, but it had no obvious connection to the realities of waking life beyond the recognition, as Emily Dickinson expressed it in 1861, that:

A single screw of flesh
Is all that pins the soul[53]

* * *

Benjamin Blood, then, was a man out of time when he began to pursue his anaesthetic revelation in the 1860s. The heyday of nitrous oxide shows and ether frolics was long past; the medical miracle of gas and vapours had eclipsed their metaphysical mysteries. Yet the stubborn questions posed by the experience remained unanswered. Blood sent copies of his pamphlet to every eminent figure he could think of, but received few responses. One surprising exception was Alfred, Lord Tennyson, who had recently had an operation and replied that 'the friend who held my hand and supplied the handkerchief' soaked in ether had told him that, as he returned to consciousness, he 'blurted out a long metaphysical term which he could not reword for me'.[54] Tennyson extended a generous but impractical invitation for Blood to visit him, and Blood filed his letter in the ever-expanding dossier of his readers' and correspondents' ineffable experiences.

'What is it? What does it all mean, or amount to?'[55] The questions posed by surgical patients nagged at William James. As he had written in his 1874 review of the pamphlet, the 'interpretation of the phenom-enon Mr Blood describes is yet deficient'.[56] In James's intellectual circles, anaesthetic visions may have been dismissed as meaningless glitches in a disconnected brain, but in 1882 James was persuaded by his brief immersion in the world of pure experience to probe their sources and meaning, and link their insights back to everyday reality.

His starting point was his long-standing struggle to make sense of the philosophy of Hegel. He was instinctively repelled by Hegelian dogma: its pompous insistence on absolute truth, its sterile abstractions, its lofty disdain for the pragmatic and the specific. James believed philosophy to be in large part determined by temperament, and he had noted *ad hominem* that those among his colleagues who were swept away by Hegel during his American vogue in the early 1880s tended to be high-minded prigs. In the paper he was writing, 'On Some Hegelisms', he spelled out why he was not a Hegelian: conflicts and contradictions could not be dismissed simply with hand-waving appeals to a higher synthesis.

In the postscript written after his nitrous oxide experiment, however, he declared that the gas had 'made me understand better than ever before both the strength and the weaknesses of Hegel's philosophy'.[57] He could now find liberation in the Hegelian scheme in which nothing was fixed or final, and everything a phase or aspect of a larger whole, to which it must always be seen in relation. Conflicts and contradictions, from this elevated view, were ultimately a matter of perspective – and that made the experience of nitrous oxide as valid as any other. Hegel's limitations were, James concluded, largely temperamental: he was determined to insist on the absolute truth of his system with a rigidity that made a nonsense of it. But within the radically altered world produced by the gas, absolute truth was itself rendered relative: it could be acknowledged 'either in a laugh at the ultimate nothingness or in a mood of vertiginous amazement at a meaningless infinity'.[58]

The anaesthetic revelation had equally profound consequences for James's other professional domain, psychology. His studies were rooted in physiology and anatomy, and during his visit to Germany in 1867 he had studied with the pioneers of the new 'psychophysics' and learned of the laboratory experiments that Wilhelm Wundt and Hermann von Helmholtz were using to measure mental processes with unprecedented precision. He witnessed thoughts and reactions reduced to their constituent parts and assembled into tables of sensory stimuli and responses, with the gaps between them measured to the split second. Concepts such as 'attention' were analysed by exposing experimental subjects to flashing lights and asking them to press keys in response, generating a precise timing for the 'reflex arc' of response, which could be mapped against observations of nerve cells and fibres.

James followed this experimental work closely, but his experience on the gas reinforced a growing sense that the attempt to reduce consciousness to its constituent parts and locate it in the brain was fundamentally flawed. He had a small experimental laboratory himself at Harvard, but found it of limited value. Mental experience, in his view, was a chaotic, ever-changing but inseparable whole, no individual moment of which could ever be precisely reproduced, even when prompted with the exact same stimuli. Subjective, variable and inscrutable processes determined the imprint of each element on the conscious mind. 'Attention', for example, was not simply a physiological reflex but a function by which the mind pulled one strand out of a mass of thoughts, feelings and sense-impressions, often in ways of which the subject was not consciously aware. 'What is called our "experience"', he wrote, 'is almost entirely determined by our habits of attention': we fail to notice the vast majority of what enters our field of perception, yet a single unusual observation may remain lodged in our minds for life.[59] Consciousness was not a linear chain of discrete thoughts or 'atoms of feeling' that could be plucked out of the brain, mapped and tabulated;[60] it was, as he would later write, more like a river or a stream: 'Let us call it the stream of thought, or of consciousness, or of subjective life.'[61] The stream of consciousness, as it would be known, was doubtless reflected in some 'cerebral action' in the brain that corresponded to each of its swirling elements, but these units of attention or perception were ultimately an artefact of scientific measurement, not of thought itself.[62]

If philosophy was a question of temperament, James's attitude to mystical experience was coloured by his impossibly complicated relationship with his father. His exasperation with Hegel's dogmatic absolutes stood as a proxy for those held rigidly by Henry James Sr, whose conflicts with his own dogmatic father had set him on a stubbornly idiosyncratic course. He had been disinherited from vast family wealth and was swept up in the Second Great Awakening, born again as an evangelical Christian before suffering a nervous collapse – a 'vastation', as he thereafter called it – and converting to the doctrines of Emmanuel Swedenborg, a Swedish engineer who, in 1744, had been visited by divine visions and gifted (as he and his followers believed) with clairvoyant sight. Throughout William's childhood his father was a prolific writer of prophetic tracts, filled with Platonic ideals and absolute truths, which struggled to find readers. He was the patriarch of a household in constant upheaval, shifting between New York, Boston and Europe; William's upbringing was varied and stimulating, and he had learned French and German on his foreign travels, but it had left him rootless and insecure.

As the eldest son, William was presented with a fiendish set of double binds. He admired his father and strove to respect his beliefs and moral prescriptions, but could not escape the fact that he found them absurd. He dreaded being forced into the role of nay-sayer and rationalist scoffer, but was incapable of playing the part of anointed son. Unable to square his respect for his father's vigour and authority with his crank religiosity, he was nagged by doubt about his own reasonableness. As a young man he was mentally fragile and had a nervous collapse at around the same age as his father had; he was diagnosed with neurasthenia, and struggled for years to find a suitable vocation or career path. He had never experienced a religious ecstasy such as his father's, nor the spiritual vastation that left him so vulnerable despite his power and charisma. He feared he had missed out on a crucial aspect of human experience that would leave his philosophy forever half-formed. Despite his physical weakness he sought out extreme and mystical experiences: his nitrous oxide experiment, which some of his Harvard students and colleagues considered reckless, was part of a wider pattern of spiritual exploration that encompassed yoga, spiritual healing and innumerable séances.

His nitrous oxide epiphany was more than a metaphysical or psychological breakthrough: it was a profound emotional release. Mysticism was no longer a foreign language to him. He could appreciate for the first time how a religious vision such as his father's could be more real than the everyday reality they shared. 'One cannot criticise the vision of a mystic', as he would later put it;[63] but by the same token, one person's mystical visions had no sovereign authority over the reality experienced by others. Both his insights and his father's convictions were equally valid. Under the influence of the gas it might be possible to map the borderlands of the mind scientifically, and believe the impossible at the same time.

*   *   *

James's journey to the world of pure experience lent urgency to his project of constructing a new psychology that could encompass mystical experience. As the mind sciences narrowed to exclude phenomena that could not be captured in objective terms by external measurement, subjective experience was being pushed to the margins: Humphry Davy's expansive, poetic embrace of a world composed of thoughts had been overwritten, along with his Romantic conception of energies, fluxes and vital forces. For the modern generation of scientists, the anaesthetic revelation was, as John Collis Warren maintained, merely a drunken delirium, and its claims to mystical insight were, as Oliver Wendell Holmes's wry anecdote implied, trivial and nonsensical. Private mental events – those without an external cause, imperceptible to other observers – were viewed through the prism of the word that now denoted such experiences: hallucination.

The term was of surprisingly recent origin, coined in 1817 by Jean-Étienne Dominique Esquirol of the Salpêtrière Hospital in Paris, the dominant figure in what had been known as alienism but was by the 1880s, in Germany in particular, being referred to as psychiatry. Esquirol's stated goals were relatively modest: to create a single category for all sense disturbances. Previous terms such as 'vision', 'apparition' or 'spectral illusion' were all weighted towards the visual, and he wanted a word that applied equally to hearing voices, feeling an invisible presence, or the sensation of bugs under the skin. He also wished to make what he conceived as a primary distinction between misperceptions of actual phenomena, which he termed 'illusions', and perceptions constructed entirely by the mind. Interpreting a fleeting shadow as a person or hearing a voice in a babbling stream were illusions; but in hallucinations, 'everything takes place in the brain; visionaries and ecstatics are hallucinators, they are dreamers who are wide awake'.[64]

By William James's time, Esquirol's description of the hallucinator as a 'visionary' sounded outdated. It harked back to the language that had been prevalent before his intervention, when private sensory events might have been understood as oracles, messages from ancestors or spirits, or the voice of God. The Italian word that Esquirol repurposed, *alucinari*, signified a wandering mind, adrift of its moorings: Dante, most famously, had used it to describe the effects of the siren song on Odysseus. But from the late 1830s, when 'hallucination' was adopted in clinical and then everyday language, it began to reshape the meaning of private experience. Visionary states had traditionally been described in terms of 'soul' and 'spirit', aspects of the self that could receive perceptions or messages from a supernatural source; but the term 'hallucination' carried the implicit judgement that these were not messages from beyond but errors in mental functioning. This made them by definition pathological and, as diagnoses of mental illness expanded and asylums proliferated, they were increasingly viewed as symptoms of insanity.

In mid-century French medicine, there was fierce debate about whether hallucinations could coexist with reason. For the positivists, the two were opposites. Some physicians, such as Alfred Maury, physician, member of the Institut de France and author of *Sleep and Dreams* (1861), had no hesitation in stating a direct equivalence between hallucinators and the insane: 'for what are the latter, if not minds who believe in their hallucinations as if they were serious facts?'[65] Maury classified dreams as 'hypnagogic hallucinations', and regarded them as mental detritus that rose up when the rational mind was disengaged. In 'retrospective medicine', a term coined in 1869 by the physician and positivist philosopher Emile Littré, the same argument was extended to the religious visions of the past: Littré and his colleagues diagnosed Moses, Socrates and Muhammad as epileptics, hysterics or paranoiacs.

Other French physicians, often those more sympathetic to Catholicism, rejected these materialist and anti-clerical diagnoses. Alexandre Brière de Boismont, an asylum doctor in Montmartre, argued in his book *Hallucinations* (1845) that historical saints and visionaries had expressed the prevailing ideas of the time, and had often harnessed them to projects for reforming society and humanity. Visions of this kind had only become pathological in the modern age, where they were alienated from the social consensus. Yet this interpretation, as much as that of the secular progressives, left no scope for understanding modern mystical experiences outside the ambit of mental disease. As the medical judgement on hallucinations hardened into orthodoxy, such experiences became less visible in normal life, and in particular to those consulting doctors who they feared might commit them to an asylum. The mysterious phenomenon of pain from amputated limbs, for example, had been anecdotally familiar at least since it was described in Descartes's *Meditations*; but before the American neurologist Silas Weir Mitchell could offer the first full account of 'phantom limb syndrome' in 1872, he had been obliged to spend many years delicately eliciting case histories from invalid soldiers who had kept their strange sensations to themselves for fear of being locked up.

By the 1880s, the growth of mental institutions was at its high-water mark across the United States and Europe. State, county and private asylums filled with patients as fast as they opened, but their rate of recovery and cure remained stubbornly low. The medical advances that were transforming surgery, pathology and infectious disease remained elusive for disturbances of the mind. Psychiatrists and asylum doctors theorised that many mental diseases were incurable: permanent damage to the nerves or brain brought on by heredity, bad habits and moral weakness. The field was claimed by physicians such as Henry Maudsley, who abandoned asylum-doctoring to edit the influential *Journal of Mental Science*, and whose 1878 paper 'Hallucinations of the Senses' drew a firm line against any attempt to indulge them, let alone induce them. There were all too many people, Maudsley wrote, 'suffering under some form or other of nervous disorder', who were convinced of the reality of their hallucinations. ' "You assure me", they will say, "that I am mistaken" ', but they remained unshakeable.

What are we to reply? I have replied sometimes, 'that as you are alone on one side in your opinion, and all the world is on the other side, I must needs think, either that you are an extraordinary genius, far in advance of the rest of the world, or that you are a madman a long way behind it; and as I don't think you to be a genius I am bound to conclude that your senses are disordered.' But the argument does not produce the least effect.[66]

Hallucinations, Maudsley wrote, had two general causes: either an overexcited mental state or a physical impairment. Anaesthetics, along with other mind-altering drugs, belonged in the second category, alongside 'fevers and some other bodily diseases' and 'exhaustion of the nerve-centres', or neurasthenia, resulting from factors such as fatigue, starvation or congenital weakness.[67] Joan of Arc's visions were, he presumed, products of the former category – 'the enthusiasm of a mind which was in a singularly exalted strain of religious and patriotic feeling'[68] – and the Prophet Muhammad's likely of the latter (epilepsy). Maudsley conceded that such visions had made a large impression on history, to the extent that 'one is almost driven to despair' by the weak-mindedness of humanity,[69] but he assured his readers that good mental hygiene could keep them to a minimum. He recommended the cultivation of healthy mental habits, physical sobriety, and the avoidance of introspection in favour of maintaining 'habitual contact with realities in thought and deed'.[70]

This was a challenging climate in which to make the case for the anaesthetic revelation, but James was convinced it was possible to establish a middle path between a medical science that dismissed his experience as delusional and a religious faith that claimed its own revelation as absolute truth. In many respects, drugs were the most promising avenue for studying such phenomena in an age of scientific measurement. Unlike spontaneous mystical experiences, drug experiments were testable and repeatable: doses could be precisely controlled, variables excluded and physiological responses studied under controlled conditions. Even if the phenomena under investigation remained ineffable and unverifiable, their physical correlates could be observed and subjective accounts from a large cohort of experimental subjects could be assembled and compared.

In 1882, just as James made his nitrous oxide experiment, a learned society was inaugurated in London to advance precisely this kind of research. On 20 February the Society for Psychical Research convened for the first time, with the intention of studying apparitions, mystical experiences, hauntings, thought transference, life after death and similar subjects that lay, in the words of its co-founder, the classical scholar and poet Frederick Myers, 'on or outside the boundaries of recognised science'.[71] In December 1882 James's father died and he left Boston to stay with his brother Henry in London, where he spent convivial evenings at the Royal Society and the Philosophical Club with the coterie of British philosophers and psychologists who were studying anomalous experiences and extraordinary mental states. The anaesthetic gases and vapours, having claimed their exalted place in physical medicine, were embarking on another journey, into the hidden depths of the mind – or, perhaps, beyond it.

CHAPTER FOUR

THE UNSEEN REGION

James's introduction to the newly constituted Society for Psychical Research, or SPR, came through the musician and essayist Edmund Gurney, an acquaintance of his brother Henry, who invited him to dine with his philosophy discussion club. Gurney was also studying hallucinations, and coming to very different conclusions from Henry Maudsley. In his view, such experiences were not pathological but normal; after all, 'are not dreams by far the most familiar instances of the projections by the mind of images that are mistaken for realities?' Hallucinations had different properties – for example, dreams tended to fade on waking while hallucinations retained their sense of reality – but he conceived them both on a continuum that included drug-induced visions. Indeed, 'the waking-dreams of haschisch-poisoning', he theorised, might form 'a sort of intermediate link' between them.[1] Gurney's particular interest was in the possibility of projecting hallucinations to others in altered mental states such as hypnosis, which he suspected might be the means by which mediums operated in séances. He chaired the SPR's Committee on Mesmerism, as well as taking on the post of honorary secretary.

Alongside Gurney in the wheelhouse of the new society were two of his old friends from Cambridge University, Frederick Myers and Henry Sidgwick, who was now the senior professor of philosophy at Cambridge. Though they had dedicated themselves to amassing scientific evidence for the reality of mystical experience, none of the three was a wholehearted believer in the scientific worldview. Henry Sidgwick was agnostic and open-minded, but unwilling to accept a materialist view of the mind that dismissed visionary experience as illusory. He was committed to establishing the reality of life after death and saw the SPR as a support network for those, like himself, collecting evidence for it. His interest was more philosophical than scientific: he believed that if the prospect of a future life was removed, the moral underpinnings of society would be eroded.

Frederick Myers was a seeker across the domains of science and religion, both of which he believed were too narrowly constituted to give such experiences their rightful place. His journey had led him from Platonism to Darwinism, with a brief conversion to evangelical Christianity and an enduring adherence to the metaphysical poetry of William Wordsworth. His view of visionary experience was theorised in psychological terms with which James felt a strong kinship. Myers saw consciousness as a spectrum of different mental frequencies, a similar view to James's stream of consciousness. 'I accord no primacy to my ordinary waking self,' he wrote, 'except that among my potential selves this one has shown itself the fittest to meet the needs of common life.'[2]

Hallucinations, in Myers's scheme, were 'uprushes' from dimensions of the mind beneath the level of conscious awareness. Myers referred to this hidden mind as the 'subliminal self'; he believed that it manifested in pathologies such as hysteria, but was also experienced routinely in dreams and creative inspiration, and could generate the insights and visions attributed to genius. It might erupt in shocking or distressing forms as a result of hidden trauma, but it might equally serve beneficial functions, for example in healing, invisibly orchestrating 'mind cures' in cases of nervous illness. It could be explored with techniques such as hypnosis, mediumship or automatic writing; he researched these phenomena assiduously, attending over a thousand séances. He was particularly attentive to the possibility that the subliminal self might project itself beyond the body into other minds, a power for which he had recently coined the term 'telepathy'.

James shared the SPR's doubts about the methods and strictures of materialist psychology, and entertained their speculations in a spirit that ranged between enthusiasm and ambivalence. He kept the door open to supernatural possibilities while careful to note that their existence was not yet proven: 'the next twenty-five years', he wrote, would 'probably decide the question. Either a flood of confirmatory phenomena, caught in the act, will pour in . . . or it will not pour in.'[3] The SPR was dedicated to amassing such evidence, and Gurney and Myers had begun collaborating on a landmark survey of suggestive cases – from predictive dreams to thought transference between twins, clairvoyant trances to the visions of mystics – that emerged in 1886 as *Phantasms of the Living*.

In theory, drugs were a royal road to the subliminal self. Unlike mystical experiences or spirit apparitions, they could be summoned on demand and repeated at will. Myers hypothesised that different classes of drugs functioned as proxies for different levels of the subliminal mind: 'the successive discoveries of intoxicants, narcotics proper, and anaesthetics', he wrote in 1886, 'formed three important stages in our growing control over the nervous system'.[4] In practice, however, they raised uncomfortable questions, both practical and ethical. 'Can we really make this an experimental study?' he wondered. 'Is it possible to induce hallucinations? And, if possible, is it safe?' Drugs were well suited to the objective demands of laboratory studies, but their effects were nonetheless unpredictable. 'The action of drugs in generating hallucinatory sounds and images is at present ill-understood,' Myers observed; on the single occasion he himself had experimented with hashish, he had simply fallen asleep. Despite their potential, he decided to err on the side of caution: 'I do not recommend my readers to take drugs for the purpose of inducing hallucinations.'[5]

Some of the SPR's members, naturally, were familiar with anaesthetics from dentistry or surgery. Edmund Gurney had had a tantalising glimpse of the beyond in the dentist's chair, but by the time he had returned home and attempted to write it up for William James, 'somehow the letter did not seem it would have much in it and it didn't get written'.[6] The member with the greatest experience was William Ramsay, the Scottish chemist who would win a Nobel Prize in 1904 for his work on isolating and identifying the gases argon, helium, krypton, xenon and neon. Ramsay had previously served on a British Medical Association committee that investigated anaesthetic gases and vapours, during the course of which he took them himself 'at least fifty times',[7] and he reported on these experiences to a packed meeting of the SPR at Westminster Town Hall in June 1893. Ether, chloroform and nitrous oxide, he told them, had different qualities but they all produced essentially 'the same mental state', which he characterised as a 'curious delusion'.[8] At first, disjointed impressions crowded in on him, competing for attention with his assigned tasks of answering questions and responding to stimuli such as flashing electric lights. As he went deeper under the influence, however, the external world receded and:

> An overwhelming impression forced itself upon me that the state in which I was then, was reality; that now I had reached the true solution of the secret of the universe, in understanding the secret of my own mind; that all outside objects were merely passing reflections on the eternal mirror of my mind . . .[9]

Each time he underwent the experience, he found, he was 'able to penetrate a little further into the unfathomable mystery'. He read Humphry Davy's famous report, and studied the idealist philosophy of Bishop Berkeley. He became more familiar with the stages of the inner journey, and the sensations that accompanied them. On each occasion he felt a more powerful certainty that 'I *know* the truth of Berkeley's theory of existence', that nothing exists but thoughts. And yet each time he was struck more forcefully that 'it is not satisfying to realise that the whole goal of the universe is of this nature'. The conviction that 'all fellow-creatures are products of my consciousness' was solipsistic and ultimately meaningless:

14. Sir William Ramsay, who won a Nobel Prize for gas chemistry, reported to the Psychical Research Society in 1893 on the disembodied consciousness induced by inhaling nitrous oxide, ether and chloroform.

My feelings are sometimes those of despair at finding the secret of existence so little worthy of regard. It is as if the veil that hides whence we come, what we are, and what will become of us, were suddenly rent, and as if a glimpse of the Absolute burst upon us. The conviction of its truth is overwhelming, but is painful in the extreme. I have exclaimed, 'Good heavens! Is this all?'[10]

Ramsay ended his lecture by highlighting a paradox in using drugs for psychical research. As material substances, they were ideally suited to the experimental method of science: they could be given in controlled doses to multiple subjects on repeated occasions, allowing for far more robust evidence-gathering than the unpredictable scenarios of séances, dreams or spontaneous mystical experience. Yet their materiality brought the authenticity of their mystical effects into question, suggesting they might be simply a physiological response wrung from a disordered or impaired brain, or a puzzle created by the temporary absence of a conscious self. 'I believe that an examination of such a mental state is not without value', he told the assembled Society, but 'I must leave to others speculations regarding the nature of the change which has occurred in my brain in such stimulation.'[11] Benjamin Blood, who had sent his pamphlet to most of the SPR's members, read Ramsay's paper eagerly but was disappointed by his conclusions. 'Sir William's depression under the commonplace and secular tone of the world-mystery', he wrote, was misplaced. His 'disillusion is in the fact that the revelation makes so intensely secular, inevitable and homely what he had before regarded as necessarily sacred and imposing and foreign'.[12]

Others, however, did succeed in using anaesthetics to stimulate spiritual and clairvoyant activity. The Scottish physician George Wyld was a member both of the SPR and of the British National Association of Spiritualists, an organisation with similar aims but more congenial to the community of spirit mediums and their audiences.[13] Wyld wrote orthodox textbooks on diseases of the heart, liver and lungs but he was also a practitioner at the London Homoeopathic Hospital and an early adopter of mesmerism. In 1874 he wrote in *The Lancet* about the strange twilight states he had witnessed, in patients and in himself, on recovering from chloroform anaesthesia. He discovered these while suffering from a painful kidney stone and inhaling chloroform for relief, at which point he:

suddenly, to my surprise, found my ego, or soul, or reasoning faculty, clothed, and in the form of my body, standing about two yards outside of my body, and contemplating that body as it lay motionless on the bed.[14]

Astonished by this brief taste of disembodied consciousness, Wyld began asking other physicians and surgeons whether they had witnessed or experienced any similar experiences. The reports came readily. One, on chloroform, had 'found himself, as it were, pleasantly whirling and soaring in the air'. A surgeon told him that 'my patients have often said that under my operations they felt no pain, but *saw* all I was doing like spectators looking on and watching the operations'.[15] Others, after reading Wyld's report, chipped in with their own stories. Some were the familiar tantalising glimpses: one, after anaesthetising a patient, 'an eminent literary reviewer and critic', was told that 'I thought I had, in some way, you know, got to the bottom and behind everything', and 'understood the great mystery that all have sought . . . but I now remember nothing more than this'.[16] But others offered apparently concrete evidence for their psychic journeys. Wyld was particularly struck by the case of a woman from Kirkcaldy who, after being administered chloroform during childbirth, returned to consciousness with the news that she had seen her mother and her baby, both dead and together in the spirit world. It was only on recovering that she learned her baby had died during the birth, and several hours later that the assembled company learned her mother had also passed away earlier that day.

This was precisely the kind of evidence that Gurney and Myers were compiling in their *Phantasms of the Living*, and Wyld believed that it demonstrated the reality not just of a subliminal self but of a soul capable of astral travel to other planes. In 1880 he became president of the British Theosophical Society, though his enthusiasm waned after he met Madame Blavatsky: he found her 'coarse and rude', reminiscent of 'a worn-out actress from some suburban theatre in Paris', and he left the Society in 1882.[17] In 1880 he published a book, *Theosophy and the Higher Life, or, Spiritual Dynamics and the Divine and Miraculous Man*, in which he considered the mechanism and meaning of these experiences. 'Did the chloroform produce clairvoyant lucidity?' he asked.[18] The cases he had been collecting were, he suggested, the most effective route to validating such phenomena scientifically:

> The sceptic will deny that the all but universal belief of human beings in the existence of the soul has any scientific weight. He will further deny the authority of spiritual revelations. He will discredit the experiments of mesmerists, and deny the assertions of Hindu or Christian ecstatics: but if he experiment with medical anaesthetics on his own person, and find out, as I and others have done, that the soul may be projected outside the body, and externally exist as the true ego, he may then be induced to believe in the existence of the human soul.[19]

Wyld's theory was that anaesthetics eliminated physical pain by driving the soul out of the body – a process that also occurred in sleep, hypnosis, trance, visions and near-death experiences, where 'the carnal state' is 'on the ebb, and the spiritual on the flow'.[20] He supported his view with Emmanuel Swedenborg's doctrine that 'Every man has an inferior or exterior mind, and a mind interior or superior'.[21] The interior mind was normally hidden from its rational possessor, but its reality was at the heart of religious traditions across the globe and throughout history, and particularly in the example of Jesus Christ.

Unlike most of his fellow physicians and SPR members, Wyld had no qualms about recommending the experience far and wide: 'I would therefore urge on Scientists, Psychologists and Materialists further experiments with anaesthetics as a means of arriving at an experimental demonstration of the existence and powers of the human soul.'[22] His explanation was adopted by many of his fellow spiritualists and echoed by the author, political activist and champion of Theosophy, Annie

Besant, in her popular work on its teachings, *The Ancient Wisdom* (1897), where she wrote that 'anaesthetics drive out the greater part of the etheric double, so that consciousness cannot affect or be affected by the dense body'.[23] The effect of these drugs, in Besant's view, was analogous to the practice of mediumship, in which the etheric double became detached from the medium's body and was able to animate extraneous objects: producing apports from thin air, moving planchettes and even taking on physical existence in forms such as ectoplasm.

Esoterically inclined surgical patients also provided vivid accounts of astral travel. The Irish playwright J.M. Synge had an operation to remove an enlarged gland on his neck, at the time when his discovery of Spinoza was leading him on to the Theosophical and occult literature that framed the last, mystically oriented decade of his life. In his account *Under Ether* (1897) he wrote that before the procedure he had lain in bed reading Spinoza, 'the great pantheist';[24] the following morning, in the operating theatre, as the ether-breathing apparatus was placed over his nose and mouth, he sensed 'clouds of luminous mist were swirling about me'. 'I'm an initiated mystic!' he heard himself yelling in sudden fury, and at that moment 'my life seemed to go out in one spiral yell to the unknown':

> The next period I remember but vaguely. I seemed to traverse whole epochs of desolation and bliss. All secrets were open before me, as simple as the universe to its God. Now and then something recalled my physical life, and I smiled at what seemed a moment of sickly infancy . . . These earthly recollections were few and faint, for the rest I was in raptures I have no powers to translate.[25]

Eventually the clouds rolled over him again, and when they parted he found himself staring at the gaslight suspended above the operating table. His vision receded: 'Oh, if I could only remember!' he groaned, as the nurses bustled about him. He 'took notice of every familiar occurrence as if it were something I had come back to from a distant country'.[26] In later years he credited his vision, or astral journey, with attuning him to 'the profound mysteries of life'.[27] He concluded his essay: 'The sentiment was very strong on me that I had died the preceding day and come to life again, and this impression has never changed.'[28]

*   *   *

By the 1880s the term 'psyche' was undergoing a pronounced shift of meaning. Where it had previously tended to imply soul or spirit, it was now used – as, for example, in the new disciplines of psychology and psychiatry – to denote the mind, which in turn was presumed to be ultimately reducible to the brain. Spiritual phenomena were nonetheless of great value to mental science, just as scientific acceptance was to the psychical researchers. Many of the new generation of psychologists, including Sigmund Freud and Carl Jung, were drawn to studying séances and mediums. Their interest was not in validating supernatural claims or exposing fraud, but in studying psychic phenomena as manifestations – along with somnambulism, automatic writing and fugue states – of a subconscious, hidden or separate mind. The subliminal self and the etheric double were, from this perspective, not two opposed theories but two formulations among many of the uncanny phenomenon known as 'double consciousness', a subject of fascination for scientists, spiritualists and public alike.

The signature tool for investigating double consciousness was hypnosis, a technique that had been reclaimed from the discredited theories of mesmerism by researchers such as the Scottish physician James Braid, who had shown by self-experiment that it could be induced by simple techniques such as concentrating on a fixed point or object. How hypnosis worked was vigorously disputed, but it seemed to inhibit the critical, controlling self, or ego, allowing access to parts of the mind beyond or behind conscious awareness. French psychiatry led the field, in the wake of its celebrated practitioner Jean-Martin Charcot, whose work with hysterical patients at the Salpêtrière Hospital in Paris had shown that that, under hypnosis, they could repeatedly manifest not only bizarre neurological symptoms but in some cases entirely separate personalities. By the 1880s several of Charcot's students, including Pierre Janet, Alfred Binet and Charles Richet, had taken the procedure out of the clinic and into the laboratory. Using hypnosis on volunteer subjects, together with instruments such as metronomes, they measured attention, suggestibility and recall, and used experiment to tease apart the conscious and the unconscious aspects of the mind.

This opened up to empirical scrutiny a range of twilight states such as trance possession and mediumship, in which the personality seemed to be ousted from the body by another entity, whether subconscious or spirit. In the course of his research, Pierre Janet developed the concept of 'automatisms' – actions of which the dominant personality was unaware – and coined the term 'dissociation' to describe the splitting of consciousness into two entities unaware of each other. It had long been known that some individuals appeared to house two separate personalities, often with no knowledge or memory of each other. Sometimes the phenomenon seemed to have its root in an accident, head injury or emotional trauma, but it had been most commonly observed as a temporary effect of anaesthetic drugs. From Humphry Davy's experiments onwards, on public stages and in operating theatres alike, an air-bag of nitrous oxide or an ether-soaked cloth had conjured up thousands of subliminal selves and provoked extravagant antics of which, on recovering, their subject had no recall.

Charcot had hypothesised that only hysterical subjects were susceptible of being hypnotised, but the prevalence of automatisms and dissociation under anaesthetic drugs buttressed the view that, as Charles Richet put it, everyone might unwittingly possess 'an unconscious ego, an unconscious activity, which is constantly on the watch, which contemplates, which gives attention, which reflects, which forms inferences, and lastly which performs acts – all unknown to the conscious ego'.[29] 'Is it not extraordinary', wrote Alfred Binet in *On Double Consciousness: Experimental Psychological Studies* (1890), that 'there should exist two distinct personalities, two egos united in the same person?'[30]

In 1885 Frederick Myers led an SPR delegation to France to witness Pierre Janet's investigations into hypnosis and telepathy, and Charles Richet participated in similar experiments under the SPR's auspices in London. But if double consciousness created bonds between the spiritualists and the scientists, it also split the new discipline of psychology into opposing camps. For the increasingly dominant cadre of laboratory psychologists, introspective journeys into unconscious realms were a retrograde step. Wilhelm Wundt dismissed the pursuit of what he called an 'aimless chase after wonders'.[31] Spiritualist séances and supernatural phenomena, for him and his followers, were a remnant of the bygone era of superstitions, ghosts and witches. They regarded hypnosis in such cases merely as mesmerism in scientific garb, producing a tapestry of lurid and self-aggrandising fantasies.

The mysteries of double consciousness resonated far beyond the internecine disputes of psychology. In popular culture, drugs and hypnosis took on the potent and sinister identity of agents that could unleash hidden personalities, create doppelgängers or conjure mind-controlling entities. Popular fictions of possession and haunting that were previously regarded as supernatural were reinterpreted in the light of modern science. In George du Maurier's *Trilby* (1894), often reckoned to be the bestselling novel of the Victorian era, the evil hypnotist Svengali transformed the title character into two 'mutually amnestic' personalities, as the psychological jargon had it: Trilby, when not under the hypnotist's spell, had no recall of her existence as a famous singer. Dr Jekyll and Mr Hyde were, by the same jargon, 'mutually cognizant' dual personalities, each aware of the other's existence. Robert Louis Stevenson's novel, its personality-switching drug positioned at the centre of the mystery, was often cited in scientific debates on double consciousness. It prompted questions about fiction-writing itself: were novelists, whose characters often seemed to appear and perform autonomously in their heads, adepts at channelling the subliminal contents of their minds? Charles Dickens had famously conversed with his characters as he went about his daily business, and Stevenson was a particularly suggestive case: he often stated that 'Brownies' or 'little people' came to him in his dreams with ideas for his stories, and William James was fascinated to learn that he had four distinctly different styles of handwriting.[32]

The young H.G. Wells, as ever a barometer for the scientific interests of his day, wrote a string of short stories in the 1890s that revolved around double consciousness and its attendant mysteries: the relation between memory and identity, the definition of personality, the limits of suggestion and mental control. In 'The Remarkable Case of Davison's Eyes' (1895), a laboratory accident leaves the title character with a consciousness split between his vision, which gazes out on an unknown and desolate stretch of ocean, and his sightless but otherwise unaltered body, stumbling round the laboratory of his London technical college. In 'The Story of the Late Mr Elvesham' (1896), a mysterious powder allows the mind of a wealthy dying man to take possession of the body of a young and healthy host. In 'The Stolen Body' (1898), a psychical researcher succeeds in achieving astral projection only for a malign spirit to enter his vacated body. In 'Under the Knife' (1897), the protagonist is chloroformed on the operating table and finds himself, as in George Wyld's reports, hovering above his body, watching as the surgeons cut him 'like cheese'.[33] As if tugged by a magnet, his consciousness rushes upwards, revealing first the West End of London, then the landscape of southern England, then the contours of the planet, and beyond the solar system the blackness of space, the Milky Way receding, and finally an intimation of the anaesthetic revelation, the outlines of which were by now familiar to his readers:

> Was the whole universe but a refracting speck upon some greater Being? Were our worlds but the atoms of another universe, and those again of another, and so on through an endless progression? And what was I? Was I indeed immaterial? . . . And then a voice, which seemed to run to the uttermost parts of space, spoke, saying, 'There will be no more pain.'[34]

*　*　*

15. *Illustration for H.G. Wells's 1897 short story 'Under the Knife', in which chloroform sends a surgical patient on a disembodied cosmic journey.*

By this time ether and chloroform were no longer confined to operating theatres. They were widely available in pharmacies, inhaled as soothing vapours for chest and lung conditions, as an analgesic for aches and pains, and as fast-acting tranquillisers for panic attacks and other nervous conditions. From the late 1880s, William James used chloroform regularly for his insomniac nights when he got 'out of the state of torpor into that of jiggle';[35] he frequently oscillated between overstimulation and exhaustion and, like Freud, diagnosed himself as neurasthenic – his 'old weakness', as he ruefully described it when 'very much run down in nervous force'.[36] Patent syrups and cough remedies typically included a measure of ether or chloroform to provide a cooling sensation as they evaporated in the throat, as well as a dose of opium or morphine to suppress the cough reflex.

In pharmacy products, chloroform was preferred for its smell – sweeter and more medicinal than ether's sharp industrial odour – and its higher evaporation point that made it less volatile and flammable, a significant advantage in an era of wooden interiors and often leaky gas lighting. Although it was usually inhaled in domestic settings, some nervous sufferers carried it with them in public as a discreet remedy for anxiety attacks. Doctors and journalists commented disapprovingly on its 'luxurious' use in tea rooms and theatres, and on the occasional public sightings of groups of young women giggling and swooning under its influence.

At the same time it developed a sinister reputation, thanks to a handful of sensational criminal cases such as that of Henry Howard Holmes, who used it in the murders of an unknown number of people in Chicago during the World's Fair of 1893. In the popular true-crime magazines, chloroform accrued the instantly recognisable whiff of malfeasance and, usually, murder. In legal cases, its discovery at the crime scene plunged basic facts into doubt: the testimony of witnesses might be impaired by blackouts, or the phenomenon of double consciousness might mean that those who had committed crimes were mentally absent and therefore legally innocent. By the late 1890s it was conjoined in yellow journalism and the popular imagination with addiction, suicide, rape and murder, and with the enduring misconception that a chloroform-soaked rag held over a victim's face produces an instant loss of consciousness (in reality, this requires continued deep breathing).

Adding to the fear and confusion was the fact that, in rare cases, it had been found to cause instant death by stopping the heart. One tragic fatality was the SPR's Edmund Gurney, who was found dead in a Brighton hotel in 1888 with an open bottle of chloroform by his bed. Gurney suffered from excruciating facial neuralgia, for which he had been prescribed many drugs including chloral hydrate, belladonna and morphine. The coroner's verdict was accidental death, though as often in such cases there were rumours of suicide. Gurney's hotel room door had been locked on the inside, and a waterproof sponge bag placed over his nose and mouth.[37]

Against this background, it was important for scientific researchers to signal that their self-experiments with anaesthetics were of a quite different order from intoxicating or 'luxurious' use by the non-medical public. James Crichton-Browne, medical director of the West Riding Lunatic Asylum in Yorkshire, made a series of experiments on his patients, and also on himself, with drugs including morphine, cannabis, chloral hydrate, ergot, ether and nitrous oxide. In this last case, published in *The Lancet* in 1872, he invoked the Cartesianism of the genteel, maintaining that the meaningful aspects of the experience were only available to scientifically trained observers. The effects of the gas, he wrote,

> in persons of average mental calibre . . . are pleasant and stimulating, but in no way remarkable; but in persons of superior mental power, they become thrilling and apocalyptic. A working man who inhales the gas intimates on his recovery that he felt very happy, as if he had had a little too much beer, and a philosopher announces that the secret of the universe had been, for one rapt moment, made plain to him.[38]

One of the fullest and most insightful contributions to the self-experimental literature came from the unlikely source of Benjamin Ward Richardson, whose unimpeachable reputation was enough to protect him from criticism. Richardson was a physician, lecturer and prolific author, best known for his firm advocacy of total abstinence: he believed that alcohol should be prohibited for all but emergency medical uses. Richardson was alerted in his capacity as a sanitary inspector to an outbreak of ether-drinking in Draperstown in the Sperrin Mountains of Northern Ireland, where vigorous temperance activism and a hike in excise duty rises had combined to make ether cheaper and more available than alcohol. Alighting from his coach in the market square, the tell-tale vapours hit him 'as certainly as if I had been in the sick room using spray for an operation'.[39]

Richardson noticed that many of the traders in the market smelled of ether, and 'passing along so that the wind brought the vapour from the lower part of town, I easily traced the odour of the vapour several hundred yards' to the door of the house where it was being sold.[40] He spent several days compiling a full report, in which he detailed who was using ether and how: remarkably, the heaviest users had hit upon a method of drinking it rather than inhaling it, by chasing a shot with a glass of water to keep it from evaporating back up the throat. He wrote up his experience in a much-syndicated long essay on 'extra-alcoholic intoxication' that first appeared in *Popular Science Monthly* in 1878, in which he distinguished carefully between the effects of methylated and ethyl ether and described the effects of the latter on himself. He took as his charter Humphry Davy's 'memorable, perfect and original work' on nitrous oxide, and presented his own impressions in the same scientific and heroic spirit:

It seemed to me that the space of the small room in which I sat was extended into a space that could not be measured and yet could be grasped and threaded; as if my powers, mental and physical, adapted themselves, for that moment, to the vastness of that space. It seemed to me that every sense was exalted in perceptive appreciation. The light was brilliant beyond expression, yet not oppressive; the ticking of a clock was like the musical clang of a cymbal with an echo; and, things touched felt as if some interposing gentle current moved between them and the fingers . . . They who have felt this condition, who have lived, as it were, in another life, however transitory, are easily led to declare with Davy that 'Nothing exists but thoughts! – the universe is composed of impressions, ideas, pleasures and pains!'[41]

* * *

The strands of medicine, consciousness expansion, intoxication, addiction and crime were tightly entangled in *fin-de-siècle* Paris, where ether and chloroform circulated among bohemian *demi-mondaines* alongside morphine, opium, cocaine, hashish and wormwood-infused absinthe. They were often carried in small glass vials and medicine bottles by the asthmatic, tubercular and neurasthenic, added to patent tonics and syrups and, on occasion, to cocktails: an ether-soaked strawberry floating in champagne produced a heady rush, the fruit preventing the volatile liquid from evaporating too quickly. Literary references to ether abounded, either as a signifier of decadence or as a literary prop to shift a realistic narrative into the landscape of dreams and symbols, where its dissociative qualities became a portal to strange mental states, psychological hauntings, uncanny doublings and slippages of space and time.

Ether was the subject of Guy de Maupassant's 1882 short story 'Dreams', in which a group of jaded and neurasthenic diners lament the *ennui* that consumes their days and the insomnia and bad dreams that ruin their nights. The doctor among them assures them that 'real dreaming', which is 'the sweetest experience in the world', is in the gift of modern medicine. His fellow diners assume he is referring to opium or hashish, which they have already tried; one replies wearily: 'I have read Baudelaire, and I even tasted the famous drug, which made me very sick.' But the doctor is speaking of ether, which he first tried to soothe his neuralgia, 'and which I have since then, perhaps, slightly abused'. Lying down with an ether bottle and inhaling slowly, he recalls, his body:

became light, as light as if the flesh and the bones had been melted and only the skin was left, the skin necessary to enable me to realize the sweetness of living, of bathing in this sensation of well-being. Then I perceived that I was no longer suffering. The pain had gone, melted away, evaporated. And I heard voices, four voices, two dialogues, without understanding what was said.

Unlike hashish dreams or the 'somewhat sickly' visions of opium, the state that followed was one of heightened mental clarity:

I reasoned with the utmost clearness and depth, with extraordinary energy and intellectual pleasure, with a singular intoxication arising from this separation of my mental faculties . . . It seemed to me that I had tasted the Tree of Knowledge, that all the mysteries were unveiled, so much did I find myself under the sway of a new, strange and irrefutable logic. And arguments, reasonings, proofs rose up in a heap before my brain only to be immediately displaced by some stronger proof, reasoning, argument. My head had, in fact, become a battleground of ideas. I was a superior being, armed with invincible intelligence, and I experienced a huge delight at the manifestation of my power.[42]

The doctor continued inhaling from his bottle for an eternity, carried away in his cerebral reveries, until he looked down and saw that it was empty. Maupassant's own ether use combined the medical, the sensual and the philosophical in similar ways. He first tried it as a remedy for his catalogue of persistent medical and neurological conditions, which included migraines, rheumatism, partial blindness, internal bleeding and fevers; his physicians offered conflicting opinions on their cause, most of which have been retrospectively connected to his eventual deterioration and death from syphilis. The regular use of ether affected him strangely: he described to friends seeing little red men sitting in armchairs, feeling his soul separating from his body and, more than once, walking into his house and seeing himself sitting on his sofa.[43]

Through the 1890s, as his condition worsened, Maupassant's hallucinations and psychotic episodes grew more intense, whether from degenerative brain disease or the overuse of ether and other drugs. Ether's imprint is discernible in the auditory hallucinations he describes, which commence as a ringing in the ears, mount towards a crescendo, and resolve during the ensuing reverie into voices, their words often inaudible or nonsensical but the character and tone of the speaker sharply defined. Disembodied voices were often described in both the medical and the spiritual literature, and were experienced and interpreted in various ways. The British doctor Ernest Dunbar described hearing them on chloroform in a report to the SPR:

> It seemed to me that deep down somewhere in my consciousness voices were wrangling and quarrelling . . . This sort of conversation would commence: 'So you see we've got you again'. Then I would think, 'Oh! won't you leave me alone? I want to rest', and the answer would come, 'We'll have the last word', and with that would ensue a muttering and a grumbling, which occasionally rose to a whining complaint from these voices. I have never inhaled chloroform without hearing these voices.[44]

Where psychic researchers heard telepathic communications, the spirits of the deceased or the voices of angels, Maupassant tended to interpret them as doublings or splittings of his own mind. In his final years, however, he was unable to dismiss the possibility that they were intimations of something beyond the self: not necessarily spirits or demons as traditionally conceived, but evidence for a disembodied presence that was stalking the modern world. His most sustained fictional exploration of this possibility was his 1887 short story 'The Horla', which abandons the elegant naturalism of his earlier work for a fractured, ambivalent and perverse narrative in which the rug is constantly pulled from under the narrator's feet. In a series of ever more harrowing diary entries, the protagonist documents a double consciousness or self-haunting, in which objects in his locked room are interfered with. 'I must be the plaything of my enervated imagination', he decides; or perhaps he has become an amnesiac or a sleepwalker; 'or I have been brought under the power of one of those influences – hypnotic suggestion, for example – which are known to exist, but have hitherto been inexplicable'.[45] But if he is mad, how can he be so clear-headed and rational?

> Some unknown disturbance must have been excited in my brain, one of those disturbances which physiologists of the present day try to note and fix precisely, and that disturbance must have caused a profound gulf in my mind and in the order and logic of my ideas.[46]

As under ether, arguments and proofs construct themselves with ever greater ingenuity, only to be replaced by others. The anomalies persist and intensify, the gulf widens; before long he is speculating that 'Somebody possesses my soul and governs it!'[47] The idea takes hold that the scientific investigators of double consciousness have unleashed an entity that feeds on the conscious mind, and over which they have no control:

> Mesmer divined him, and ten years ago physicians accurately discovered the nature of his power, even before He exercised it himself. They played with that weapon of their new Lord, the sway of a mysterious will over the human soul, which had become enslaved. They call it mesmerism, hypnotism, suggestion . . . A new being! Why not? It was assuredly bound to come! Why should we be the last?[48]

The literary figure most closely associated with ether in *fin-de-siècle* France was the novelist, poet, journalist and short-story writer Jean Lorrain, a collection of whose dark and sardonic tales was published in

*16. The French decadent author Jean Lorrain, a habitual user of
ether, whose fiction explored its uncanny psychic effects.*

1895 under the title *Nightmares of an Ether-Drinker*. Lorrain was an
extreme figure even in decadent Paris: a powdered and bejewelled
dandy, scandalous gossip-monger, bohemian, dabbling Satanist,
denizen of the city's criminalised and violent gay underworld, and at
the same time its most highly paid writer. 'What is a vice?' he shrugged.
'Merely a taste you don't share.'[49] His brazen self-advertisement as the
Thomas De Quincey of ether was entirely in character, although
the stories in the volume tend to mention the drug only obliquely and

he never describes his experiences with it as directly as Maupassant did in 'Dreams'. Its presence is diffuse, tearing at the veil of reality and punctuating it with surreal juxtapositions and flashes of preternatural clarity. It sets the scene and the mood for stories such as 'An Uncanny Crime', which its narrator begins:

> It was two years ago, when my nervous troubles were at their worst. I had recovered from the ether, but not from the morbid phenomena to which it had given birth: hearing things, seeing things, nocturnal panic-attacks and nightmares. Sulphonal and bromide had begun to alleviate the worst symptoms, but my distress continued in spite of medication. These phenomena were at their worst in the apartment on the other side of the river, in the Rue St Guillaume, which I had shared with them for so long. Their presence seemed to have impregnated the walls and fittings, by means of some pernicious sympathetic magic . . . There were bizarre shadows huddling in the corners, suggestive folds in the curtains at the windows, while the door curtains would suddenly be animated by some frightful and nameless semblance of life.[50]

Lorrain first used ether medicinally, as symptomatic relief for his chronic tuberculosis – similar to his rival Maupassant, who once challenged him to a duel for plagiarism.[51] His illness, and his self-medication with ether, were both woven into his highly cultivated public persona. A close friend of Joris-Karl Huysmans, whose *À Rebours* (1884) established itself as the decadents' bible, Lorrain's writing career straddled the worlds of high aestheticism, well-paid journalism, commercial pulp sensation and unprintable perversion. Like Huysmans' masterpiece, many of his stories are not so much narratives as vignettes, mood pieces or studies of mental states, unfolding in interior monologues that anticipate the modernists' stream of consciousness. *À Rebours* includes a lengthy digression on scent, and the transports and associations that can be summoned by it; ether in the writing of

Maupassant and particularly Lorrain can be seen as an outgrowth of this obsession amplified to madness. In the salons and cafés, it was often said that you could smell Lorrain's presence in the room before you saw him. In his fiction, its odour and its dream logic pervade throughout, bringing with them a confusion of associations: the hospital waiting room, the boudoir, the agony of the lungs, the gentle dissociation from reality, the sudden awakening from a nightmare – and, ultimately, the pain of the gastric ulcers that his habit inflicted on him with growing severity.

Although his ether tales are profoundly haunted, Lorrain was no believer in ghosts or spirits; as he might have said, quoting Samuel Taylor Coleridge, 'I have seen far too many myself.'[52] Like Huysmans, he investigated Satanism: the pair were habitués of the Chat Noir café, Montmartre's bohemian hub where the disciples of Jules Michelet discussed witchcraft and sought out secret rites in basements and catacombs, and rumours of secret covens and occult rituals swirl in the background of his tales. In the end, however, neither science nor the supernatural, nor indeed ether, offered sufficient explanation. 'Oh, don't blame it on the ether!' replies the protagonist in the short story 'The Possessed' to a friend trying to make sense of his terrifying obsessions:

> Nevertheless, I do have to go away. I'd be sure to fall ill again as soon as November arrives, when Paris becomes fantastically haunted. You see, the strangeness of my case is that I no longer fear the invisible, I'm terrified by reality.[53]

Lorrain's perverse and shocking novel *Monsieur de Phocas* (1901) marked the scandalous high-water mark of his career, and the *ne plus ultra* of French decadent literature. Bankrupted by lawsuits for plagiarism and obscenity, he fell from fashion in the new century and in 1906 suffered an ignominious death of peritonitis after perforating his colon with an enema while attempting to relieve his ether-induced intestinal ulcers. The drug had, like De Quincey's opium, become his talisman, his trademark and his curse. It opened the door to an invisible dimension of mind that hypnotism and the mind-doctors were no more capable of explaining than mediums or exorcists. As another of his overwrought protagonists exclaims in the 1891 vignette 'Magic Lantern': 'Never has the Fantastic flourished, so sinister and so terrifying, as in modern life!'[54]

*   *   *

Twenty years after he published his first impressions of nitrous oxide in *Mind*, William James placed his experience on the gas at the centre of his magisterial survey of mystical states, *The Varieties of Religious Experience* (1902). 'Nitrous oxide and ether, especially nitrous oxide,' he wrote, 'stimulate the mystical consciousness in an extraordinary degree. Depth beyond depth of truth seems revealed to the inhaler,'[55] he continued, reproducing almost verbatim his original insight that 'Truth lies open to the view in depth beyond depth of almost blinding evidence.'[56] His mature reflections culminated in his best-remembered paean to the drug experience:

> Some years ago I myself made some observations on this aspect of nitrous oxide intoxication, and reported them in print. One conclusion was forced upon my mind at that time, and my impression of its truth has ever since remained unshaken. It is that our normal waking consciousness, rational consciousness as we call it, is but one special type of consciousness, whilst all about it, parted from it by the filmiest of screens, there lie potential forms of consciousness entirely different . . . No account of the universe in its totality can be final which leaves these other forms of consciousness quite disregarded.[57]

James had still not foreclosed on the nature or meaning of mystical experiences, but after immersing himself in hundreds of descriptions, from the diaries of medieval mystics to the confessional accounts of the

Great Awakening to the piles of personal correspondence that now filled his study, he had sketched its contours. The failure to describe it satisfactorily, he concluded, was not proof of its incoherence, but of its authenticity. Such experiences were ineffable, beyond words: 'its quality must be directly experienced; it cannot be imparted or transferred to others'. They were also defined by what he called a 'noetic' quality, carrying with them the unshakeable conviction of truth: 'they are illuminations, revelations, full of significance and importance, all inarticulate though they remain'.[58] They were transient, slipping from conscious apprehension almost immediately; and they were passive, impossible to achieve by exercise of will, which gave them the sense that they had been gifted by a superior power. They were 'absolutely authoritative over the individuals to whom they come', as they had been to his father, but they could not give their subjects authority over others, as his father had demanded. They implied a further truth, or synthesis, that applied equally to both sides: that 'they open out the possibility of other orders of truth'.[59]

The 'unseen region' was the capacious term that James now used to describe the source of experiences that might be claimed, or dismissed, as religious or mystical, pathological or drug-induced. Among his examples were several cases from the files of the Society for Psychical Research, including spontaneous hallucinations, automatic writing, spirit apparitions and presences; but his commitment to the SPR's cause was waning. In 1894 he had accepted its presidency on Frederick Myers's insistence; Myers had convinced him that his international fame and status made him the obvious candidate, but he resigned the following year in favour of the distinguished chemist Sir William Crookes, inventor of the cathode ray tube. The twenty-five years James had originally proposed as the time-frame for their project was almost up, and the world was no closer to accepting the reality of telepathy or life after death. On the contrary, it had moved firmly in the opposite direction: his perceived spiritualist leanings had dented his prestige, and his influence on other causes to which he was equally committed.[60] 'I am theoretically no "further" than I was at the beginning', he wrote as he looked back on his psychical researches in 1909; 'and I confess at times I have been tempted to believe that the Creator has intended this department of nature to be *baffling*, to prompt our curiosities and hopes and suspicions in equal measure.'[61]

James was more opposed to reductive and dogmatic science than ever, but he now framed his approach to the unseen region differently. Where the SPR were interested in their cases primarily as evidence for the reality of another plane of existence, James argued for their reality on their own terms. The stream of consciousness was a 'blooming, buzzing confusion' of many different modes of thought and perception, where deep currents of mystical experience often mingled with the surface froth of daily life.[62] Whether we regard these experiences as true or delusional, they impact our real lives: 'work is actually done upon our finite personality, for we are turned into new men, and consequences in the way of conduct follow in the natural world'.[63]

The anaesthetic revelation might generate claims that fly in the face of science and logic, but that did not make them necessarily supernatural, nor necessarily profound. Pierre Janet had come to a similar conclusion after his extensive hypnotic investigation of spirit mediums: their prophecies and pronouncements, though sometimes remarkable and baffling, were mostly perfunctory, predictable or banal. Yet James's own experience had taught him that extraordinary mental states could not be assessed on the basis of the testimony that arose from them; and, given the abundant evidence he had amassed for the transformative power of visions, he concluded: 'I feel as if we had no philosophic excuse for calling the unseen or mystical world unreal.'[64]

Through James and many other sources, the model of mind and reality sparked by the anaesthetic revelation flowed into the wider stream of influences that redefined consciousness in the twentieth century. The modernist art and culture of the generation to come, with its fractured and unfamiliar perspectives, dedicated itself to breaking through the veil of everyday reality to a world of pure experience. The ether-induced mindscapes of Jean Lorrain, haunted by their masked figures and doppelgängers, looked forward to the interior monologues of Marcel Proust and James Joyce, the uncanny stream-of-consciousness narratives of Arthur Schnitzler and the impossible juxtapositions of the surrealists. During the 1880s, artists such as Georges Seurat began to experiment with divisionism and pointillism, shifting the locus of vision from the external world to the eye and mind of the beholder; Paul Cézanne and the Cubists gave visual expression to James's worldview, combining multiple perspectives of the same object in a single canvas. The rigid boundaries of time and space were loosened by the new technologies of phonograph and cinema that froze transient moments and perceptions into material objects. When Wassily Kandinsky broke through into pure abstract art in 1910 – 'painting the soul', as he conceived it[65] – he drew inspiration from the unseen regions he visualised through meditating on Theosophical thought-forms.

In science, too, the collapse of time and space experienced under anaesthetics was extended from the subjective to the objective and measurable world. William James proposed in his *Principles of Psychology* that subjective time perception, unlike clock time, was elastic, reflecting the frequency of mental events that were occurring: 'awareness of *change* is thus the condition on which our perception of time's flow depends'.[66] He connected this to the thinking of Henri Bergson, whom he regarded as the great philosopher of direct experience, who argued that time resides in the body as memories do in the brain, as a lived and unmeasurable dimension of consciousness, and the rigid increments of clock time are at best a simplification of this reality. When Albert Einstein's formulation of spacetime emerged in the 1900s, it presented an even more profound collapse of objectivity: time was not a single and universal measure but perspectival and plural, ultimately dependent on the relation between objects. In 1910 the Spanish philosopher José Ortega y Gasset advanced the insights of Einstein and others to the conclusion that 'There is no absolute space because there is no absolute perspective . . . the theory of Einstein is a marvellous proof of the harmonious multiplicity of all possible points of view.'[67] In 1916 Einstein concluded that 'every reference body has its own particular time'.[68] As James insisted after his revelation on nitrous oxide, no account of the universe can be final without including the face that it presents to each unique observer.

A similar insight was transforming the social sciences. In 1888 the anthropologist Franz Boas had been the first to use the term 'culture' in the plural, to reflect that 'civilisation is not something absolute, but that it is relative, and that our ideas and conceptions are true only so far as our civilisation goes': beyond the European inheritance lie other cultures, and forms of consciousness, entirely different.[69] The same year, the twenty-year-old W.E.B. Du Bois, the path-breaking civil rights activist and Pan-Africanist of the generation to come, entered Harvard and found his great teacher in William James, 'my friend and guide to clear thinking', as he later wrote.[70] Du Bois became the first Black student to earn a Harvard PhD and, in 1895, in an article in the *Atlantic Monthly*, he adapted the term 'double consciousness' to describe the situation of Black Americans, obliged to navigate the world simultaneously through their own perspective and that of the dominant white culture which set its rules and parameters. 'It is a peculiar sensation, this double-consciousness,' he wrote, 'this sense of always looking at one's self through the eyes of others.'[71] The path to Black identity in the modern world involved reconciling 'vast and partially contradictory streams of thought'[72] into a higher synthesis, that of 'the sovereign human soul'.[73]

* * *

The culmination of James's journey was spelled out in his essay 'A World of Pure Experience' (1904) and a final lecture series – to the largest ever crowd for a philosophy lecture at Harvard – published in 1909 as a book, *A Pluralistic Universe*. He described his philosophy as 'radical empiricism', which he defined as 'the opposite of rationalism': instead of privileging reason and abstraction, it insisted on direct experience and the individual fact. Reality, in this scheme, was the sum total of all experience, whether material or unseen, and nothing more; it 'must neither admit into its constructions any element that is not directly experienced, nor exclude from them any element that is directly experienced'.[74] It rejected equally the strictures of logic and reason, the revealed truth of religion and the Absolutes of Plato or Hegel. It was a mosaic in which every piece could be fitted together to produce a coherent version of the world, but the image they created was dependent on how they were assembled and where the observer was situated. The word James coined to denote the totality of the mosaic was the 'multiverse'. It was a final and decisive break with the laboratory psychology that attempted to isolate the functions of the mind and find the basis of consciousness in the physiology of the brain and nervous system. Consciousness and indeed reality, James asserted, were ultimately snapshots of a process that was constantly being shaped by their experiencing subject.

The last piece of writing that James published in his lifetime was a short appreciation, entitled 'A Pluralist Mystic', of the figure who had been instrumental in setting him along this path:

> Now for years my own taste, literary as well as philosophic, has been exquisitely titillated by a writer, the name of whom I think must be unknown to the readers of his article; so I no longer continue silent about the merits of BENJAMIN PAUL BLOOD.[75]

Blood, James recalled, had released him from the unsatisfactory choice of responding to mystics either by affirming their claims or rejecting them. His anaesthetic revelation had been an Absolute, but one that contained all the contradictions of identity and difference, mind and matter, thesis and antithesis in its multiverse, all resolved in its 'trumpet blast of oracular mysticism'.[76] Blood had often written of 'the One', and sometimes identified it with God, but in his later writings he shifted towards James's view: he titled his own final book *A Pluralist Universe*. There was only experience; as for the unseen region beyond it, as Blood had written in a letter to James after his nitrous oxide experiment of 1882, the ultimate truth is '*we do not know!*'[77]

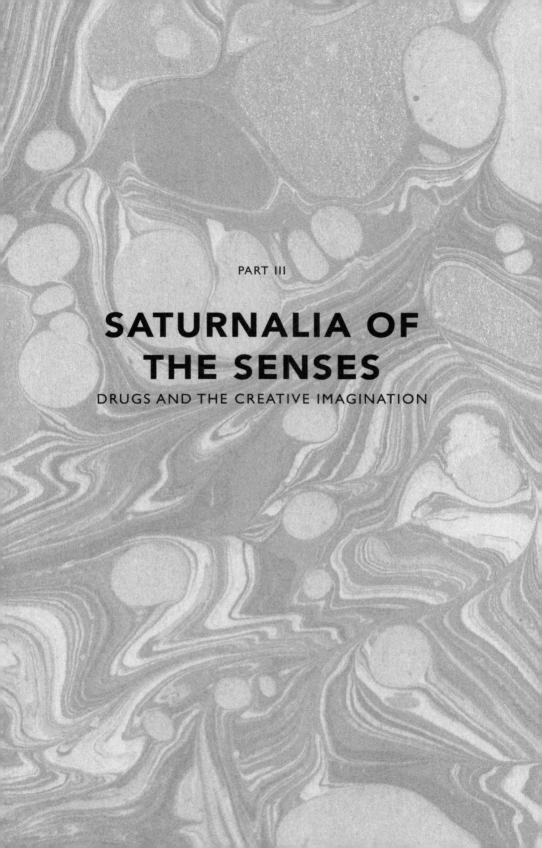

PART III

SATURNALIA OF
THE SENSES

DRUGS AND THE CREATIVE IMAGINATION

17. A drawing under the influence of hashish by Jean-Martin Charcot as a young medical student, c. 1850.

CHAPTER FIVE

TALES OF THE HASHISH EATERS

In April 1892 the British trade journal *Chemist and Druggist* carried a report of travels and adventures in Morocco from the manufacturing chemist and pharmacy magnate Silas Burroughs, co-founder of the innovative and hugely successful Burroughs Wellcome & Company. Burroughs had stopped in a Moorish café in Tangier for coffee (and green peppermint tea for his guide) and decided to experiment with the cannabis that was also on offer. A boy cut up some leaves from the Indian hemp plant and mixed them with tobacco, and Burroughs took 'a few whiffs' from the long wooden pipe. Everything 'grew bright' around him, and he was 'instantly carried off into the seventh heaven of imaginary delight'. His description, narrated to the *Chemist and Druggist* reporter, reads like a fantasia from the pages of *The Arabian Nights*:

> You are transported, like Aladdin, to the moonlit Alhambra, and hear the music of its fountains and streams, or down the Red Sea and beyond the pinnacles of Sinai to gaze at distant Mecca, and then you are floated as if by an invisible balloon over the snowy peaks of the Himalayas, to the evergreen foothills where sacred streams burst from beneath marble temples and flow by the heathen shrines.[1]

Silas Burroughs was, on the face of it, an unlikely psychonaut. He was a staunch Presbyterian and strict teetotaller, a fitness enthusiast and energetic pioneer of the bicycle. He was also, however, a self-described 'inveterate globetrotter', a pioneer of recreational travel who, at the height of his career success, absented himself from running Burroughs Wellcome to spend months of every year in foreign parts, developing international business opportunities while seeking out strange and exotic experiences.[2] He found cannabis delightful: it makes you 'feel so light', he declared, 'that you could run up the jagged sides of the pyramids'.[3] His experience fulfilled the sensual and mysterious promise of Morocco's souks, where pharmacists' stalls were stocked from floor to ceiling with jars, bottles and drawers of curious preparations, local herbs such as sarsaparilla and pennyroyal jostling with amulets and charms against the evil eye and European patent medicines such as effervescent liver salts. A staunch supporter of progressive social movements, unionised labour and tariff-free trade, Burroughs was pleased to discover that the apothecary was among the few classes of Moroccan businessman not obliged to pay tributes to the Sultan on imported goods. 'Perhaps,' he speculated, it was because 'he brings the means of recovery to health to so many of the faithful.'[4]

Cannabis was one of many medicines that was available in both the bazaars of the Orient and the modern pharmacies of the West, yet Burroughs's account makes no claims for its medical potential. In Europe and the United States a generation earlier, cannabis extracts and tinctures had been hailed as one of the wonders of modern pharmacy. By the 1890s, however, the plant's medical promise had faded: it hung on as a minor ingredient in patent sedatives and was recommended by some doctors for muscle cramps. Commercial pharmacy had moved on, and cannabis was ill-suited to the new market. Unlike the coca leaf and the opium poppy, which had yielded the pure drugs cocaine, morphine and codeine, it defied all attempts to reduce it to a standardised chemical compound. It was typically supplied to pharmacists in the form of hashish, a concentrated plant extract imported from Egypt or India that varied from batch to batch, making it unsuitable for 'ethical' or 'scientific' pharmaceuticals that listed their active ingredients in precise doses.

Silas Burroughs was at the forefront of the modern trends. During the 1880s, Burroughs Wellcome had begun to develop 'compressed medicines', replacing the pharmacists' traditional jars of powders weighed at the counter into paper wraps with coloured and distinctively branded pills produced by prototype tableting machines. In 1884, having found that the British had an aversion to the term 'pills', Burroughs Wellcome trademarked the word 'tabloid' which, along with other terms such as 'elixoid', they used to market the pure products that now dominated the brightly illuminated palaces of pharmacy. These products included several powerful mind-altering drugs. Their range of soluble-compressed hypodermic tabloids included pure preparations of morphine, along with strychnine and atropine; in 1885 they produced a 'Voice Tabloid' for public speaking, whose chief active ingredient was cocaine, which boosted confidence as effectively as it cleared the airways. In 1890 they launched 'Forced March', an energy-boosting pill containing cocaine and caffeine that remained one of their biggest sellers until it was taken off the shelves at the start of the Great War.

Despite his public stance against intoxication – he would shortly join the National Temperance League and sport its blue ribbon in his button-hole – Burroughs was happy to relate his cannabis experience to the *Chemist and Druggist's* readers without any pretence to medical research. He presented it, rather, as exotic ethnography. As he told the reporter, he had not taken a dose of hashish – the medical preparation familiar in Europe – but smoked a pipe of *kif*, a preparation of finely chopped leaves and stalks. This method of smoking was long established in Morocco and the Maghreb, and had recently been taken up in the cosmopolitan *demi-mondes* of some Western cities; but elsewhere, the drug was traditionally eaten. Hashish, the resinous concentrate made from the flowering heads of the cannabis plant across the Arab world and the Indian subcontinent, was often mixed with fruit, nuts, rosewater and syrup to counteract its bitterness, creating a sweet confection known in Egypt as *dawamesc* and Morocco as *majoun*.

These oral preparations had long been consumed in doses that produced powerful and unpredictable effects. Often their subjects would remain prostrated for several hours, unable to move or speak, immersed in visions and fantasies that succeeded each other too rapidly and chaotically to describe. Others would become maniacal, filled with irrepressible energy or gripped by obsessions and delusions incomprehensible to those around them. In their spectrum, intensity and duration, these effects were comparable to what we now call psychedelic drugs; and when the major psychedelics made their debut in Western culture at the end of the nineteenth century, it was in the milieu of hashish-eating bohemians, aesthetes and occultists that they were most fully explored.

Silas Burroughs's experience was one among hundreds of descriptions of cannabis (usually hashish) intoxication that echoed through the literary and aesthetic world of the late nineteenth century. By this time the plant had accrued a reputation as a portal to mysterious and otherwise inaccessible dimensions of mind, and was being explored in many circles: scientific, literary, musical, philosophical, spiritual, occult. Like the anaesthetics, it expanded and heightened consciousness; but whereas nitrous oxide, ether and chloroform induced a disembodied state of pure thought, the intoxication of hashish was embodied and sensual. Its inner visions were represented through an aesthetic of baroque adornment, often drawing on Oriental forms, and in the visual languages of symbolism, synaesthesia and ceremonial magic, whose practitioners turned its intoxication to astral travel and the pursuit of heightened spiritual powers. While cocaine became emblematic of heightened energy and performance, a panacea for the strenuous demands of an intensified modern world, hashish became the signifier of its opposite: voluptuous indolence, freewheeling imagination and cosmic detachment. This oceanic state of blissful emptiness was seen as foreign to the Western mind, and was captured in the Arabic term *kif*: the local Maghrebi term for Silas Burroughs's smoking mixture, but used more widely in an abstract sense to describe a state of perfect, euphoric contentment.

*   *   *

By the mid-nineteenth century, hashish was familiar to educated Westerners from two colourful and unreliable sources. The first was the stories of *The Arabian Nights*, which had achieved extraordinary popular reach during the previous century, initially in Antoine Galland's French twelve-volume translation (or interpretation, or invention)[5] of 1704–17 and subsequently in a profusion of bootlegged, embroidered and bowdlerised versions. By 1882 the tales were, according to Robert Louis Stevenson, 'more generally loved than Shakespeare', a world that 'captivates in childhood, and still delights in age'.[6]

'Hashish', in *The Arabian Nights*, was a European rendering of the Arabic *banj*, which can refer to a variety of plant intoxicants: most often cannabis, but also opium or deliriant nightshades such as datura or henbane. The drug features in tales set in the streets of Baghdad during the eighth-century golden age of Sultan Haroun al-Rashid, where it circulates among low-life beggars, petty criminals and religious subversives: typical hashish eaters, in the view of the literary and cultured elite of medieval Persia. It features in several tales, and inhabits two well-defined roles. In some it was deployed as a knockout drug by criminals and kidnappers to render their victims unconscious or amnesiac.[7] In others, such as the well-known 'Tale of the Hashish Eater', it was a comic device for provoking delusions of grandeur. In this instance a beggar, reduced to poverty by his appetite for women and luxury, eats a large lump of hashish and wanders into a bathhouse, a scene transformed before his eyes into a palace where slaves wash and perfume him and he is supplied with alluring concubines. He drifts into a blissful sleep and awakes to find himself immersed in a cattle-trough in the street, naked and with an erection, to the amusement of the crowd thronging the street. It was, in the bathetic punchline, all a dream.

The hashish tales of *The Arabian Nights* were broad, raucous comedy relief, set among and appealing to the lower strata of society for whom hashish was one of life's few affordable luxuries. Their subjects were typically the feckless poor and their vain, presumptuous attempts to gain rewards without earning them. Like Oliver Wendell Holmes's ether anecdote, they were a droll assurance to the respectable and sober-minded that drug epiphanies were always exposed in the cold light of day as delusions. Once the tales reached Europe several centuries later, however, hashish was opened to fresh interpretations: a prompt to fantastical descriptions of Oriental opulence, or an affirmation of the power of the imagination to elevate the dreamer, if only temporarily, from the gutter to the stars.

The second context in which hashish was familiar was the scholarly debate around Marco Polo's famous story of the Assassins, the mysterious and deadly sect who occupied an infamous position in European cultural memory as the Crusaders' most feared adversaries. In his account, first published around 1300, Polo described (from hearsay) a fortified palace and gardens hidden high in a remote mountain cleft in the north of Persia, presided over by a potentate known as the Old Man of the Mountain. The walled gardens were exquisitely designed and tricked out with mechanical contrivances that made the rivers run with wine, milk and honey. When the Old Man's recruits and followers arrived at the citadel they were given an intoxicating drink that sent them into a deep sleep, from which they awoke to find themselves in his gardens, surrounded by courtesans dressed in gold and silk. They enjoyed this idyll of the senses until they eventually slept again, at which point they were returned to the palace. On waking, the Old Man informed them that he had magically transported them to Paradise, and that after a life and death in his service as Assassins they would return there. The recruits swore total obedience to him, most famously demonstrated by their willingness, at the snap of the Old Man's fingers, to leap off the citadel walls and plunge to their deaths a thousand feet below.

The sinister legend of the Assassins, and the brainwashing drug with which the Old Man controlled his homicidal followers, posed many mysteries for the Orientalist scholars of the eighteenth century. Some suspected that the story referred to the Nizari Isma'ilis, a breakaway sect from the Shi'a Fatimid caliphate of Cairo, but the etymology of the name 'Assassin' remained obscure, as did the stupefying plant in the Old Man's potion, which was often translated as opium. In 1809, however, in a celebrated lecture at the Institut de France in Paris, the leading Oriental philologist Antoine Isaac Silvestre de Sacy presented a solution to both riddles simultaneously. 'Assassin', he announced, was derived from 'al-Hashishiyya', an Arabic term for the sect found in several medieval manuscripts, with the literal meaning of 'hashish eaters'. The Assassins were those who had been brainwashed by hashish into committing political murders: 'delivered', as de Sacy put it, 'through the use of hashish, to this absolute resignation to the will of their leader'.[8]

De Sacy's solution posed further riddles, not least about the nature of hashish and its connection with hemp – which was also known, following classical sources, as cannabis, from which the textile 'canvas' was derived. It was a well-recognised fibre crop, widely cultivated across northern Europe, and although the plant grew differently in Arab regions – shorter and stubbier, with large, sticky flower heads – it seemed curious that the European variety had no intoxicating qualities.[9] De Sacy's sources, mostly medieval Muslim clerics, attributed terrible effects to hashish. Its users were said to be frenzied and deluded, deprived of their reason and degraded below the level of humanity. He quoted anecdotes still in circulation that recalled the world of *The Arabian Nights*, such as one of a dissolute hashish addict who believed he was a pasha, lounging on a divan, airily discoursing on high affairs of state and condemning his enemies to imprisonment.[10]

De Sacy informed his audience that 'the Arabs use the word *kayf* [*kif*] to describe . . . voluptuous abandon' and 'delicious stupor'.[11] His source for this was the French naturalist Charles Sonnini de Manoncourt, who had observed hashish users during his travels in Egypt in the 1780s. Manoncourt commented that 'our language has no terms to express' *kif*, which he conceived as a distinctively Arab state of mind in which rational thought is erased by blissful dreams and reveries.[12] It seemed curious that the same drug could also provoke the homicidal violence demanded by the Assassins' vocation, but de Sacy resolved the apparent contradiction by theorising that the Old Man had used the 'state of ecstasy and sweet, profound illusions' induced by hashish to initiate his followers, after which they were joined irrevocably to his murderous sect and needed no further drugs to commit their infamous acts. (De Sacy's identification of the 'Assassins' as Nizari Isma'ilis has subsequently been confirmed, but his assertion that they used hashish is widely rejected. They appear to have been strict, devout and ascetic, and to have prohibited all intoxicants on pain of death.)

*　　*　　*

The first sustained interaction between the modern West and the hashish eaters of the Arab world occurred during Napoleon's invasion of Egypt in 1798, which culminated in a ban on the use of hashish by French troops in October 1800. Hashish, according to the decree by General Jacques-François Menou (Napoleon himself had already left the country) caused its users to 'lose reason and fall into a violent delirium, which often leads them to commit excesses of all kinds'.[13] No direct testimony was cited from hashish-using soldiers themselves, and it seems that Menou intended the ban to align the occupying regime with the interests of the Egyptian Sunni elite who, like many of de Sacy's sources, associated hashish with the criminal underclass, Sufi mendicants and political undesirables. *Hashishin*, as it had been in the era of the Crusades, was used as a dismissive term along the lines of 'riff-raff' or 'dregs of society', the sense in which it seems to have been applied to the Nizari Isma'ilis by their enemies.

The effects of hashish on the mind were eventually established in the time-honoured manner, by self-experiment. 'I challenge the right of anyone to discuss the effects of hashish if he is not speaking for himself,' wrote Jacques-Joseph Moreau, the physician who became synonymous with hashish in Europe's medical world.[14] Moreau's interest in the drug was prompted by his travels in Egypt, Turkey and Syria from 1837 to 1840, as a junior doctor accompanying wealthy patients from the Charenton asylum in the suburbs of Paris. He had joined the asylum staff in 1826, while completing his doctoral thesis, to work under the direction of Jean-Étienne Dominique Esquirol, the doyen of French mental therapy. Esquirol followed his own mentor, Philippe Pinel, in conceiving madness, or 'alienation', not as a religious judgement or a moral weakness but a physical condition, a disease of the brain and nerves. Like Pinel, he encouraged physicians to spend time with their patients, show them empathy and guide them towards self-observation and insight into their condition. For those who could afford it, he would often arrange rest cures and changes of scene, with a junior physician such as Moreau assigned to accompany them.

Moreau was fascinated by life in the Arab world. He followed Esquirol's advice to dress in the local style, see the sights, immerse himself in the crowds and adopt the local customs. As an alienist, he was particularly interested in differences between Arab and European mentalities, and he interrogated his guide and translator, a *dragoman* named Kelim, about his inner world. Kelim was a firm believer in *djinn*, with which he had had several terrifying encounters, and in the prophetic power of dreams. These forms of private experience would be described by many European doctors as hallucinations and diagnosed as symptoms of mental disease; yet Kelim appeared otherwise quite sane, and his beliefs were typical of a population in which mental illness was encountered far less frequently than in Europe.

Moreau's doctoral thesis had been on monomania, a diagnosis developed by Esquirol that he saw as a disease of over-focused attention leading to obsessive behaviour. By 1840 it had become the most prominent form of insanity in France, accounting for a quarter of all diagnoses. Moreau wondered if the visions or fantasies that Kelim experienced, when sanctioned by a culture that permitted indulgence in a rich inner life, might be not a sign of madness but a possible remedy for it. He was particularly intrigued by the absence of alcohol

18. A hashish market in Cairo around the time of Jacques-Joseph Moreau's residency in Egypt.

in Arab culture and the corresponding prevalence of hashish, which seemed to accompany and enhance this communal fantasy world of folklore, storytelling and dreams.

Moreau discovered that he was not the only French doctor in Egypt interested in hashish. An epidemiologist named Louis Aubert-Roche, who had been based in Egypt since 1830, was investigating it as a treatment for typhoid and bubonic plague. Aubert-Roche believed these were diseases of the nervous system that spread in unhygienic conditions, and that hashish might help to prevent. 'If this plant has intoxicating powers', he reasoned, 'it must have some medical power over the nervous system.'[15] Aubert-Roche was working in a hospital in Cairo and supplied cannabis, in the palatable form of *dawamesc*, to patients, doctors and interested medical researchers, including Moreau.

* * *

The idea of developing cannabis into a Western medicine, in the form of a tincture with a standardised dose, had recently been pioneered by France's great colonial rival. William Brooke O'Shaughnessy, a dynamic Irish physician and pharmacist working in British Bengal, was, like Aubert-Roche, initially attracted by the possibility of treating local infectious diseases, in his case cholera. Previously, British physicians had paid little attention to the plant's mind-altering properties, but O'Shaughnessy recorded them with evident interest. If he experimented on himself, he left no record of it; instead, he enlisted local subjects, including 'the proprietor of a celebrated place of resort for hemp devotees in Calcutta, who is considered the best artist in his profession'. He compiled a taxonomy of local preparations that paralleled those of Egypt, including the plant itself (*gunjah*), a drink prepared traditionally for religious ceremonies (*bhang*), a hashish-style concentrated extract (*churrus*) and the sweet confection *majoun*.[16]

In a paper delivered to the Medical and Physical Society of Calcutta in 1839, O'Shaughnessy presented his experimental findings on 'this extraordinary agent'.[17] He had begun his trials with a dog that he dosed with 'ten grains of Nepalese churrus dissolved in spirit'; it became 'stupid and sleepy', 'ate some food greedily' and 'staggered to and fro' with 'a look of utter and helpless drunkenness'. After six hours it recovered, becoming 'perfectly well and lively'.[18] He proceeded to dose fish, birds and other mammals, noting only minor effects on 'the horse, deer, monkey, goat, sheep and cow', before proceeding to human subjects: patients at Calcutta's Native Medical Hospital. On these, a dose of *churrus* produced some striking physical responses, including in one instance 'that strange and most extraordinary of all nervous conditions', catalepsy, in which the patient became a waxen statue, oblivious for an hour to all external stimuli. More common were bizarre states of mind, such as that witnessed in 'an old muscular cooly [labourer], a rheumatic malingerer', who:

> became talkative and musical, told several stories, and sang songs to a circle of highly delighted auditors, ate the dinners of two persons subscribed for him in the ward, sought also for other luxuries I can scarcely venture to allude to, and finally fell soundly asleep, and so continued until the following morning. On the noon-day visit, he expressed himself free from headache or any other unpleasant sequel, and begged hard for a repetition of the medicine, in which he was indulged for a few days, and then discharged.[19]

O'Shaughnessy's research excited 'the utmost interest' among his medical staff and students, and 'several pupils commenced experiments on themselves' with the tinctures that he was producing. Unlike Humphry Davy, O'Shaughnessy did not solicit first-person testimony from his native subjects: he followed more closely the example of Robert Hooke, relaying and vouching for their experiences rather than presenting them as direct evidence. He noted their 'vivid ideas' and 'unusual loquacity', and in one remarkable case introduced his subject by name:

> In one pupil, Dinonath Dhur, a retiring lad of excellent habits, ten drops of the tincture, equal to a quarter of a grain of the resin, induced in twenty minutes the most amusing effects I ever witnessed. A shout of laughter ushered in the symptoms, and a

transitory state of cataleptic rigidity occurred for two or three minutes. Summoned to witness the effects, I found him enacting the part of a Raja giving orders to his courtiers . . . He entered on discussions on religious, scientific, and political topics, with astonishing eloquence, and disclosed an extent of knowledge, reading, and a ready apposite wit, which those who knew him best were altogether unprepared for . . . A scene more interesting it would be difficult to imagine.[20]

Like the hashish eaters of *The Arabian Nights*, Dhur's behaviour was expansive and grandiose, but in this case there was no rude awakening: his mental brilliance while under the influence was palpable, and O'Shaughnessy was lost for an explanation. He drew suggestive comparisons with 'the effects produced by the reputed inspiration of the Delphic oracles', and speculated that 'the same kind of excitement', some intoxicating plant or vapour, might have been implicated in the oracular trance state.

The recommendations that concluded O'Shaughnessy's paper focused on his trials of cannabis in cases of tetanus. It was, he claimed, highly effective, 'in a large proportion of cases effecting a perfect cure', an outcome he attributed to its anti-spasmodic and muscle relaxant properties.[21] He concluded that the violent psychic effects of the drug mentioned in the Bengali medical and legal literature were genuine but rare, and should not discourage physicians and pharmacists from including cannabis preparations in the Western pharmacopoeia. He set out the basics of a formula for pharmacists: boiling the flowering tops of *gunjah*, dissolving it in spirit (it was not water-soluble), and evaporating it to a consistency where it could be made into pills.

*   *   *

O'Shaughnessy's work prompted a surge of medical interest in Europe. Chemists in Edinburgh and Paris began to develop pills, tinctures and standardised doses for the pharmaceutical market. But the effects of hashish on the European mind still awaited a full description. This came from Jacques-Joseph Moreau, on his return to Paris in 1840 to take up the post of resident physician at the Bicêtre asylum in the south of the city. While researching a paper on hallucinations and their treatment, he made his first self-experiment with a strong dose of the *dawamesc* that Aubert-Roche had supplied him with in Cairo, in company with two friends.[22]

Moreau's first surprise was that the supposedly delicious confection was horribly bitter to taste, and he 'swallowed it with great effort'.[23] The three sat down to eat oysters, and after a little while Moreau was stricken by sudden fits of laughter. His friends noticed the effects at the same time, and drew his attention to a lion's head which seemed to have materialised on their dinner plate. Consumed by mirth, all adjourned to the drawing room, where Moreau started to play the piano before being distracted by a vision of his brother standing on top of it. After much hilarity, confusion and 'a thousand incoherent speeches, gesticulating, shouting like all the masks I believed I saw', he stumbled into an unlit room and believed he had fallen into the well in the grounds of the Bicêtre hospital. He felt thousands of insects running through his hair, and then abruptly 'an intoxicating happiness' at a cherubic vision of his young son fluttering in a dazzling blue sky on pink and white wings. The pace of the visions accelerated, as memories, illusions, nightmares and fantasies swirled together:

> I spoke about people whom I had not seen in several years. I recalled a dinner I attended five years ago in Champagne. I saw the guests. General H. served a fish surrounded with flowers . . . I felt that I was at home, that everything I was seeing had happened in a distant time. However, the people seemed to me to be there. What was I to believe? . . . I cannot describe the thousand fantastic ideas that passed through my brain during the three hours that I was under the influence of the hashish. They seemed too bizarre to be cred-

ible. The people present questioned me from time to time and asked me if I wasn't making fun of them, since I possessed my reason in the middle of all that madness.[24]

As an alienist, Moreau was fully aware that the symptoms he was experiencing bore all the signs of madness. In his subsequent book, *Hashish and Mental Illness* (1845), he itemised them methodically: nervous excitement; distortion of time and space; obsessions or *idées fixes* (for example, a recurring conviction that he had been poisoned); irresistible impulses; illusory perceptions. 'There is not a single, elementary manifestation of mental illness', he concluded, 'that cannot be found in the mental changes caused by hashish.'[25] At the same time, however, he remained entirely rational: able to diagnose his symptoms as he experienced them, and to observe calmly as a procession of impossible phenomena marched through his mind. He described this remarkable form of double consciousness as an *état mixte*, a mixed state in which two normally separate forms of consciousness, the inner world of dreams and the waking state of reason, were able to coexist and observe one another.

Here was empirical proof that madness was at root a physiological condition that could be produced on demand by a material cause: as Moreau put it, the 'phenomenal result of a toxic substance on the nervous system'.[26] He was overwhelmed by the possibilities for understanding the mind, treating mental illness and liberating the imagination. Even as hashish plunged its subject into madness, it offered glimpses of a higher sanity. Its primary symptom, he concluded, was happiness, which – like that induced by nitrous oxide – arose without any identifiable cause beyond the drug itself: 'You feel happy; you say it; you proclaim it with exultation; you seek to explain it with all the means that are within your power; you repeat it to the point of satiation. But words fail you to say how and why you are happy.'[27] Moreau recognised Humphry Davy's account as a precursor of his own experiments: 'I would try in vain to describe it better,' he wrote.[28] For him, as for Davy,

'This is indeed very curious, and one can draw strange conclusions – this one among others, that all joy, all contentment, even though its cause is strictly mental, deeply spiritual and highly idealistic, could well be in reality a purely physical sensation.'[29]

Moreau was tempted to speculate, as Freud would on cocaine, that a chemically induced euphoria was the driver that prompted the drug's expansive effects on the imagination. At the same time, his conclusion foreshadowed William James's, in his 1884 essay 'What Is an Emotion?', that we do not cry because we are sad, but vice versa. Since emotions are mental events, we assume that they have an immaterial cause and ignore 'how much our mental life is knit up with our corporeal frame'. Mood and emotion are inseparable from physical stimuli and responses: is it possible, James asked, to imagine rage with 'no flushing of the face, no dilation of the nostrils, no clenching of the teeth, no impulse to vigorous action, but in their stead limp muscles, calm breathing and a placid face?'[30] Hashish, nitrous oxide and cocaine were quite different in their effects on the mind, but in each case the onset was marked by a rush of sensory pleasure that induced a shift in mental state, and in so doing opened up normally inaccessible dimensions of consciousness. Moreau saw the happiness of hashish, like Freud's 'healthy euphoria' on cocaine, not as a delusion or impairment but an enhancement of health. His struggle to describe it was evidence of its strangeness to the rational Western mind, though it had long been familiar elsewhere under the name of *kif.*

Moreau, like Davy, was confronted by the question of how to turn his discoveries and insights to practical use. His first thought was to experiment with hashish in the treatment of mental illness, for which as resident physician at the Bicêtre he was ideally situated. The 'feeling of gaiety and joy' it produced seemed to hold great promise for 'combatting the fixed ideas of depressives, of disrupting the chain of their ideas, of unfocusing their attention'. Its mental scattering was the opposite of monomania, and its *état mixte* opened the possibility of constructive therapy, in which hallucinatory experiences could be held up to the light of reason and generate insight into the patient's condition and its causes.[31] In practice, however, the responses were highly unpredictable. Some patients appeared to notice no effect at all from their dose of hashish; others became highly excited, sometimes with fits of hysterical laughter. In some cases this was followed by a *kif*-like state of calm and lucidity in which it proved possible to engage in therapeutic conversation, though insights were hard to retain after the drug had worn off. Other subjects, however, became confused or tired, withdrawing deeper into themselves; yet others became anxious, hostile and paranoid, developing an *idée fixe* that they had been poisoned or were being possessed. It was impossible to predict how any individual patient was likely to respond.

His therapeutic trials were, by and large, a failure. 'Were my conjectures mistaken?' Moreau asked himself. 'I am led to believe so without, however, considering the matter closed.'[32] Yet he remained convinced that hashish had much to offer the treatment of mental illness, and he conceived an alternative proposal: it should not be given to the patients, but to their doctors. 'Can we be certain', he asked, that 'we are in a condition to understand these sick people when they tell us of their observations?'[33] It was a guiding principle of Pinel's and Esquirol's treatments that the alienist should attempt, as far as possible, to walk in their patients' shoes; until now, however, it had been impossible to follow them across the border that so decisively divided them, between sanity and madness:

> Do they not speak a language that is foreign to us? How can they communicate the feelings that disturb them? . . . We see only the surfaces of things; we cannot proceed further to explore the causes, the succession of mental aberrations they describe to us. Are not mental processes, feelings or sensations impossible to know and evaluate except through oneself?[34]

Within the *état mixte* of hashish, the physician could be both participant and observer, mad and sane, dreaming and awake at the same time.

The effect was 'a kind of transformation of the self or the personality': 'Dream begins where the freedom to direct our thoughts ceases . . . The mind cannot act outside of this freedom without taking on, in some sense, a new, quite independent existence, unconnected with the preceding one. A new life follows the other and replaces it.'[35]

*　*　*

Like Humphry Davy, Moreau wanted to test his discovery on healthy minds as well as sick ones, and like Davy he sought out artistic and literary subjects whose powers of description and imagination might create a language to describe the drug's effects. In 1842 this project evolved into a salon whose fame and reputation would eclipse Moreau's medical and scientific work. It convened every few weeks in the Hôtel Pimodan on the Île St Louis, an island enclave in the middle of the Seine where the medieval cobbled streets were enveloped by night in fog and silence, a refuge from the bustling, noisy city on either bank. The proceedings unfolded in the hotel's grand salon room, decorated in the Louis XIV style with gilded cornices, a marble chimney with a golden elephant and classical murals of nymphs pursuing satyrs. The salons were discreetly organised, though a written invitation survives, addressed to the author, poet and journalist Théophile Gautier by the artist Fernand Boissard, who lived in the hotel building:

> Dear Théophile
>   On Monday next the 3rd November, hashish will be taken at my place under the supervision of Moreau and Aubert-Roche. Arrive between 5 and 6 at the latest. You will have your share of a light dinner and await the hallucination.[36]

Gautier's semi-fictionalised account, 'Le Club des Hashischins', published in the literary journal *Revue des deux mondes* in 1846, was the fullest and best-remembered description. As he admitted in later life, its

19. Théophile Gautier, author of 'Le Club des Hashischins', photographed by Nadar (Félix Tournachon).

style was typical of its day and of the *Jeunes-France* movement of which he was a member: a milieu of struggling artists and self-fashioning *littérateurs* in thrall to Victor Hugo, where 'it was the fashion to be pale and greenish-looking; to appear to be wasted by the pangs of passion and remorse; to talk sadly and fantastically about death'.[37] Gautier wrote from the perspective of a trembling neophyte, taking a carriage through the foggy winter night to the 'half-worn, gilded name of the old hotel, the gathering-place for the initiates'.[38] He raps the carved knocker, the rusty bolt turns and an old porter points the way upstairs with a skinny finger.

The first figure he encounters is a mysterious master of ceremonies, 'Doctor X', who spoons a morsel of green paste onto an elegant Japanese saucer and offers it to him with a mock-Assassin benediction: 'This will be deducted from your share in Paradise.' Moreau hosted the salon in Oriental dress he had acquired during his time in Egypt: a sketch by Gautier, drawn at lightning speed under the influence of hashish, shows the doctor from behind, playing the piano in robes and a turban: 'the musical notes are shown escaping from the keyboard in the form of rockets and capriciously corkscrewed spirals'.[39] The *dawamesc* is washed down with Turkish coffee and the initiates sit down to a feast, a beggar's banquet of meat and drink served in Venetian goblets from Flemish jugs and on flowered English crockery, no two pieces alike. The narrator's companions are a motley crew in Oriental costumes, flowing robes and long beards, armed with ancient-looking swords and daggers. As he takes his place in the outlandish scene, Gautier notices the drug's creeping derangement of his senses:

> Already some of the most fervent members were feeling the effect of the green paste; personally, I had experienced a complete transposition of taste. The water I drank seemed to me to have the flavour of the most delicious wine, the meat in my mouth was transformed into raspberry, and vice versa. I could not have distinguished a chop from a peach . . . A vague warmth stole over my limbs, and a madness, like a wave breaking on a rock and retreating to hurl itself again, reached and withdrew from my brain, which it finally engulfed. Hallucination, a strange guest, had come to stay with me.[40]

As the banquet continues, the senses become deranged to the point where physical stimuli degenerate into nonsense, and the imagination reconfigures the external world. Gautier's companions become inhuman, hybrid creatures from the canvases of Hieronymus Bosch with pupils 'big as a screech owl's' and noses extended into probosces. As soon as the

20. The salle à manger in the Hôtel Pimodan, Paris, where the Club des Hashischins gathered.

dinner is over, the cry goes up, 'To the salon!' In the gilded upstairs suite of the Hôtel Pimodan 'one inhaled the luxurious airs of times gone by'. Slowly the room fills up with fantastical figures 'such are found only in the etchings of Callot or the aquatints of Goya', a 'bizarre throng' that plunges Gautier into a torrent of grotesque visions, a Walpurgis Night reminiscent of that conjured by Hector Berlioz in his *Symphonie fantastique*.[41] Demonic forms taunt the narrator as he tries to escape and cackle as his flight is slowed to a snail's pace by an unseen force. The nightmare is dispelled by the pianist – presumably Moreau – who shatters the ambience by bringing his hands crashing down onto the keys and then resolving the discord into a heavenly melody that tames the infernal chaos. The frenzy gives way to 'an indefinable sense of well-being, an unending calm . . . I was in this blessed phase of hashish which the Orientals call "kief".'[42] At the moment when the clock indicates eleven, normal time is magically reborn and normal consciousness resumed, and the exhausted initiate finds his carriage waiting for him in the street below.

Gautier's report launched the Club des Hashischins as a literary sensation and a popular scandal. The sinister legend of the Assassins was still potent in French reactionary politics. In his *History of the Assassins*, originally published in 1818 after de Sacy's identification of the mysterious order with hashish, the Orientalist and fervent monarchist Joseph von Hammer-Purgstall had sketched a secret history in which the Assassins were the ultimate source of libertinism and revolution. Their ideology had entered Europe via the Knights Templar, who had secretly propagated it through Freemasonry; hashish was their ultimate secret weapon, a brainwashing agent that made its initiates 'able to undertake anything or everything'.[43] The Club des Hashischins was a provocation that turned the Assassin legend on its head, claiming the myth as their own and glorying in its subversion of Christian civilisation. 'Hashish has replaced champagne,' Gautier announced in *La Presse*. 'We believe we have conquered Algeria, but Algeria has conquered us.'[44] The salon's ironic self-mythologising, and the scandal it created, positioned it as a forerunner of modern drug counterculture: the jaded romantics of the July Monarchy cultivated an outlaw style of long hair and outlandish dress, and defined themselves with radical politics, all-night partying and sexual libertinism. If De Quincey was the archetype for the twentieth-century 'drug fiend', the Club des Hashischins set the template for the modern 'drug scene'.

Like many of the secret societies it satirised, the Club des Hashischins was successful in launching an enduring myth while leaving many of its mundane historical details in doubt. Its reputation as a well-organised secret society with regular meetings and grades of initiation was carefully curated rumour, drawing on and parodying the Assassin legend. By some accounts it met regularly once a month; according to others, there were only a handful of meetings in total and its active life was over by 1845. Gautier himself wrote later that he gave up hashish 'after some ten experiments', partly because of the physical ordeal involved, and partly because the experience produced diminishing returns.[45] The Hôtel Pimodan seems to have been used only for a short period, after which salons convened irregularly in smaller private rooms. Virtually every luminary of the mid-century Paris literary world has been associated with the club, and few denied it; some were doubtless glad to see their names added to the list of rumoured initiates without having to swallow several grams of bitter paste and submitting to an intense and prolonged derangement of the senses.

Honoré de Balzac was a case in point. According to Charles Baudelaire, he attended a meeting at the club in 1845 but refused his dose of the green paste, distrusting any substance that might sap his will.[46] Gautier gave a fuller account of the evening, at which he claimed to have been present: it was fascinating, he wrote, to watch 'the struggle between his almost infantile curiosity and his repugnance' as Balzac asked questions, touched and smelled the *dawamesc*, and contemplated a loss of mental control that would far exceed his experiments with caffeine. Eventually, 'love of dignity prevailed'.[47] By Balzac's own account, given in a letter of December 1845, he had in fact taken a small dose of hashish at the club, but felt no effect until after he left the salon, when he 'heard celestial voices and saw divine visions', and as he descended the hotel staircase the paintings on the wall 'took on an enchanted splendour'.[48] He wrote to Moreau later: 'You know you owe me another dose of hashish, since I didn't get my money's worth the first time around. Please be kind enough to warn me in advance of the place and time, for I want to be the theatre of a complete phenomenon, so as to judge your work properly.'[49]

By the late 1840s, the French reading public was encountering hashish on a regular basis, and learning that its influence blurred the lines between dream, reverie and waking life in ways that brought reality itself into question. Gautier's autofictional essay appeared just as Alexandre Dumas's *The Count of Monte Cristo* was approaching the climax of its epic serial publication. Its plot turned on the young adventurer Franz d'Epinay's visit to the apparently deserted Mediterranean island of Monte Cristo on a hunting expedition. There he stumbled upon a band of smugglers who blindfolded him and led him to a secret cave where their leader lived in Oriental splendour under the *nom de guerre* of 'Sinbad the Sailor'. Sinbad – whom readers might have suspected by this point to be the Count of Monte Cristo in disguise – served d'Epinay a sumptuous feast, followed by a small bowl of pungent green paste. 'What is this precious sweetmeat?' his guest asked. In reply, Sinbad told him the story of Hassan-i-Sabbah, the Old Man of the Mountain, who recruited his followers by feeding them a drug that would 'take them to Paradise', after which they would 'obey his orders like those of a God'.

D'Epinay recognises the story. 'In that case,' he exclaims, 'it's hashish.' It is, Sinbad confirms, 'the finest hashish from Alexandria', He continues:

> Nature wrestles with this divine substance because our nature is not made for joy but clings to pain. Nature must be defeated in this struggle, reality must follow dreams; and then the dream will rule, will become the master, the dream will become life and life will become a dream . . . Try some hashish, my friend! Try it![50]

Dumas echoes Moreau's idea that the happiness produced by hashish has the power to alter not only the mind but external reality, and adds the suggestion that this secret has long been known to Oriental adepts. These themes were explored in great depth by Théophile Gautier's childhood friend and fellow member of the Club des Hashischins, the poet and travel writer Gérard de Nerval. In de Nerval's chaotic and inscrutable inner world, eccentric bohemianism combined with a series of mental collapses and extensive travels in Egypt, Syria and Turkey to produce a permanent state of what he described as 'the overflowing of dreams into real life'.[51] His short stories and vignettes, and his travelogue *Journey to the Orient* (1851), included hashish tales that drew on both *The Arabian Nights* and the Assassin legends to create a world he called 'the other Egypt', haunted by the impossible and the supernatural.[52] In 'The Tale of the Caliph Hakim' (1847), the caliph, travelling his kingdom incognito, visits a low-life hashish dive frequented by Nile fishermen. A young boatman named Yousouf offers him a 'greenish paste', served with an ivory spatula:

'But this paste is hashish, if I am not mistaken,' said the stranger, pushing aside the cup in which Yousouf had put a part of the fantastic mixture, 'and hashish is forbidden.'

'Everything pleasant is forbidden,' said Yousouf, swallowing his first spoonful.

Under the influence of the drug, the caliph 'seemed a prey to extraordinary exaltation: hosts of new thoughts, unheard-of and inconceivable, traversed his soul like whirlwinds of fires'.[53] He entered a phantasmagoria in which his secret imposture was challenged by an encounter with his doppelgänger, and he found himself the only sane inhabitant of a world gone mad. The story was based on a tale told to de Nerval in Lebanon, but he wove it together with the familiar *Arabian Nights* motif of the caliph travelling in disguise, and the narrative unfolds with the uncanny twists of E.T.A. Hoffman's gothic tales. Unreliable travellers' yarns, traditional fables and the logic of madness combined to create an *état mixte*, the state that Moreau described elsewhere as 'dreaming while awake'.[54] In his posthumously published account of his madness, *Aurélia* (1855), de Nerval put it simply: 'dream is second life'.[55]

Hashish was perfectly suited to the literary form for which the German philosopher and critic Friedrich Schlegel had coined the term 'arabesque', which Edgar Allan Poe invoked in the title of his 1840 story collection, *Tales of the Grotesque and the Arabesque*. Schlegel defined it as a form of Romantic irony in which familiar tropes and fables threaded through daily life, taking unexpected turns between the comic, the sublime and the nightmarish. An encounter with hashish was the perfect incitement to this form: simultaneously playful and grotesque, allowing the author to whisk the narrative from the mundane surroundings of a

writer's study, an artist's studio or a squalid underworld den to a fantastical scene from *The Arabian Nights* or a labyrinth of terrors. Gustave
Flaubert worked on a novel entitled *La Spirale* – projected but never
completed – whose protagonist was to be a hashish-addicted painter
who suffers many adversities and ends up confined in an asylum, but
retains an ecstatic dream-life in which he is elevated, God-like, above
external reality. Flaubert was concerned that his readers would not care
about events that took place solely in a dream world, so he linked them
to reality by the device of making his protagonist's real-life actions affect
his dreams: when he acted nobly, his dream-life became more rewarding.
Hallucinations, in this scheme, became a force for moral good.

*   *   *

The Club des Hashischins was a *succès de scandale*, but it was also the
perfect cover for Moreau's scientific research. If he wished to study the
effects of hashish on the imagination, he needed to create a context in
which his subjects were able to exercise their faculty to the full. Conducting
the experiments in the sterile ambience of his hospital would have made
them self-conscious at best, and at worst tipped their imaginings towards
paranoia. The principle – rediscovered by psychedelic researchers a
century later, under the name of 'set and setting' – was well understood
by his experimental subjects. As Théophile Gautier wrote later:

> It is important to be in a tranquil frame of mind and body, to have
> on this day neither anxiety, duty, nor fixed time, and to find oneself
> in such an apartment as Baudelaire and Edgar Poe loved, a room
> furnished with poetical comfort, bizarre luxury, and mysterious
> elegance . . . without these precautions the ecstasy is likely to turn
> into a nightmare. Pleasure changes to suffering, joy to terror . . .[56]

As Sheikh of the Assassins in his Oriental robes, Moreau could
control the experiment and supervise the dosage without taking on the

role of physician; he was, as he had been in Egypt, camouflaged by his exotic dress while stories of *djinn* and magic unfolded around him. The ceremonial trappings and extravagant surroundings encouraged his volunteers to give free rein to their fantasy and embrace their temporary derangement, while the doctor observed their intoxication and presided over their safe return to sanity.

Hashish and Mental Illness was widely reviewed in France, cited by the Academy of Medicine in Paris and praised in medical journals across Europe and the USA: Moreau's discoveries, announced the *Boston Medical and Surgical Journal*, 'must be of much importance to the civilised world'.[57] His emphasis on self-experiment ran counter to the positivist turn in French mental physiology and was criticised by conservative alienists such as Alexandre Brière de Boismont, who cited de Sacy and the Assassin myth to argue that hashish produced madness, and that to indulge its pathological symptoms as Moreau recommended was a dangerous game. Further afield, his method was followed by researchers such as the Italian physician Carlo Erba, who in 1847 convened a gathering of fellow doctors in a hotel room in Milan. After lunch and cigars, Erba dispensed samples of *dawamesc* he had ordered from Alexandria, and his colleagues confirmed Moreau's observations by recording 'waves of pleasure', 'an unstoppable desire to laugh' and the feeling of being 'divided into two parts, one that reasoned and observed, and another that raved and raved'.[58]

By this time pharmacists were making some progress in developing a standardised dose of hashish. O'Shaughnessy's work on cannabis tinctures was published in French in 1847 and the following year Joseph-Bernard Gastinel, a French pharmacist with a well-established practice in Cairo, produced a highly concentrated liquid which he named *haschischine* and claimed was a distillation of the pure active principle.[59] But hashish, like nitrous oxide, struggled to find a defining medical application. Following Aubert-Roche, Gastinel claimed that his tincture was an effective remedy for cholera, which was just beginning its spread into the slums to the south of Paris; but when a full-blown epidemic struck the city in 1849, eventually killing 20,000, it became obvious that hashish had no power against it. A reaction set in against the extravagant claims made on the drug's behalf, and by 1850 its brief vogue as a medicine was waning. Nonetheless, Moreau's ideas continued to influence the emerging discipline of psychopharmacology. In the mid-twentieth century, his notion that drugs such as hashish manifested the symptoms of mental illness was recapitulated under the name 'psychotomimetic', the theory that framed early research into psychedelics.

The Club des Hashischins had lapsed by 1850, but its literary legacy was crowned in 1860 with the publication of *Artificial Paradises* (*Les Paradis artificiels*) by one of its occasional members, Charles Baudelaire. Baudelaire had lived in the Hôtel Pimodan for a while and may have attended the salons there as an observer, but only participated after it moved to the opulent apartment of his friend the novelist, playwright and celebrated dandy-about-town Eugène Roger.[60] According to Théophile Gautier, Baudelaire's acquaintance with hashish was slight: he reckoned he had tried it no more than 'once or twice' as 'a psychological experience'.[61] Baudelaire's essay, originally published in 1851 but subsequently refined and polished for a decade, reprised some of the club's familiar motifs but was more ironic and sceptical in tone, reflecting both the backlash against hashish that had occurred in the interim and a modern sensibility that distanced itself from the extravagant masquerades of the previous generation.

Gustave Flaubert, in the wake of *Madame Bovary*'s trial for obscenity in 1856, was among the first to congratulate Baudelaire on the finished version of *The Poem of Hashish*, which formed half of *Artificial Paradises* alongside *An Opium-Eater*, a translation of Thomas De Quincey accompanied by reflections on a drug of which Baudelaire had considerably greater personal experience. He had, Flaubert wrote, set out 'the beginning of a science, in a work of observation and induction', adding a steely empiricism to the Romantic style with which he was more readily associated.[62] Baudelaire began his essay with a botanical descrip-

21. Charles Baudelaire's self-portrait under the influence of hashish.

tion of the plant and a survey of its preparations and names in different countries, and proceeded to echo De Quincey by dismissing the idea that any drug could bestow supernatural, prophetic or artistic powers. Hashish induces strange visions indeed, but 'we never leave the natural dream', the inner world created by the subject's 'preoccupations, desires and vices'. Just as De Quincey had stressed that 'a man whose talk is of oxen . . . will dream about oxen',[63] Baudelaire warns: 'let the sophisticates and novices who are curious to taste these exceptional delights take heed; they will find nothing miraculous in hashish, nothing but the excessively natural'.[64] He advised his readers simply to take a teaspoon of *dawamesc* with a cup of coffee on an empty stomach, and to await patiently what will be a 'long, remarkable voyage'.[65]

Baudelaire divides the voyage into three stages. The first is an 'uneasy joy' at its early intimations: 'Incongruous connections, coincidental resemblances, interminable puns, and comic sketches provide endless delight. The demon has invaded you.'[66] A sense of vertiginous acceleration takes hold, and your mental leaps become wilder and more unruly:

> Soon the links that bind your ideas become so frail, the thread that ties your conceptions so tenuous, that only your accomplices understand you. And here again you cannot be completely certain; perhaps they only think they understand you, and the illusion is reciprocal.[67]

In the second stage, 'the hallucinations begin', although Baudelaire is careful to specify, as Moreau did, that hashish does not produce 'pure hallucination', a delusional world independent of sensory stimuli, but rather the 'disruption of the senses', in which the external world is distorted and transformed, more properly described as 'illusion'. At the peak of the experience:

> The powers of perception, of taste, sight, smell, hearing – all participate equally in this progression. The eyes pierce the infinite. The ear hears sounds that are almost imperceptible amid even the most tumultuous din . . . By gradations, external objects assume unique appearances in the endless combining and transfiguring of forms. Ideas are distorted; perceptions are confused. Sounds are clothed in colours and colours in music.[68]

Gautier wrote about the transposition of tastes at the Club des Hashischins banquets; but Baudelaire's description of the cross-wiring of different senses – taste, vision, hearing and touch – became a classic account of the phenomenon that would later be named 'synaesthesia'.

He had already evoked this sensation in his poetry, notably in 'Correspondences' (1857), in which perfumes are 'sweet like oboes, green like prairies'. Over the next decades synaesthesia became a signifier of spiritual and artistic peak experience, and the ecstasy of hashish intoxication its canonical example. But this form of transcendence, for Baudelaire, was nothing supernatural or mystical; it might prompt a creative or spiritual epiphany, but in itself it was simply the mind at play, extending its repertoire beyond the normal limits of waking life.

This peak is exhausting, and it lasts for hours; but at the moment when it feels impossible to bear any more, it devolves into the third and final phase, a mental state that only profound exhaustion can wring from the human frame:

> In this new phase, which the peoples of the East term *kief,* turbulence and tempests subside, giving way to calm and immobile beatitude, a glorious resignation. Although you have long since ceased to be your own master, this no longer disturbs you. All notion of time, all painful sensations, have vanished; or if these at times dare to appear, they do so only as transfigured by the dominant sensation, and then, in relation to their usual form, they are what poetic melancholy is to true sorrow.[69]

There is, however, one more phase to come, for Baudelaire the most significant of all. It manifests itself on 'the morrow! the terrible morrow!', when your enervated and shattered nerves 'tell you that you have played a forbidden game'.[70] With hashish, one has surrendered one's will in exchange for an artificial and counterfeit experience of pleasure, unearned and without nourishment, that 'reveals nothing to the individual but the individual himself'.[71] It is as immoral as suicide, and Baudelaire insists that the law should treat it in the same way:

> We are familiar enough with human nature to know that a man who can instantaneously obtain all of the ecstasies of heaven and

earth by swallowing a small spoonful of paste will never earn the thousandth part of them through his own labour. Can you imagine a State whose citizens all took hashish? What citizens! What soldiers! What legislators! Even in the East, where the use of hashish is so widespread, some governments have understood the need to put the drug under ban. Assuredly, it is forbidden to man . . .[72]

This is a jarring shift of tone from the rest of the essay, and took Baudelaire's early readers by surprise. Flaubert's sole criticism was that 'you have insisted too much on the *Spirit of Evil*. One can sense a leavening of Catholicism here and there. I would have preferred that you not *blame* hashish, opium and excess.'[73] Evil is equated here with surrendering the will to worldly desire, a view perhaps prompted by Baudelaire's more regular exposure to the demons of alcohol and opium, with which the self-destructive cycle he evokes was harder to escape.[74] 'It is difficult to believe', Gautier wrote drily, 'that the author of the *Flowers of Evil*, in spite of his satanical leanings, has often visited artificial paradises.'[75] Baudelaire responded vehemently that 'even were the entire nineteenth century in league against me, I would not retract', before allowing that 'I reserve the right to change my mind, however, or to contradict myself at any time'.[76]

Baudelaire's judgement might also be seen as a reaction against hashish as it appeared a decade past its heyday, with its medical promise tarnished and the subversive *jeu d'esprit* of the Club des Hashischins out of season. The modern bohemian now inhabited a different world from the fogbound medieval streets of the Île St Louis. In the wake of the 1849 cholera epidemic Paris was being transformed, the warrens of the old city cleared for Baron Hausmann's wide boulevards and grand squares. Transverse modern thoroughfares now slashed across the Seine, and gaslight illuminated the new boulevards, glowing from the plate glass windows of the shopfronts, turning the streets into a glittering stage for the figure whom Baudelaire, in his famous essay of 1863, hailed as the true artist of modern life, the *flâneur*.

The lover of universal life moves into the crowd as though into an enormous reservoir of electricity. He, the lover of life, may also be compared to a mirror as vast as the crowd; to a kaleidoscope endowed with consciousness, which with every one of its movements presents a pattern of life, in all its multiplicity.[77]

It was a prescient vision of the *ville lumière*, the city of lights in which hashish, and other novel drugs of vision and pleasure, would find a new public among the generation to come. Baudelaire was quick to recognise that the sublime experience he described would not for long remain the elite preserve of the dandy or the man of taste, but would soon become one of pleasures of the crowd. The salon, with its discreet and elitist overtones, was being replaced by a preening, exhibitionist café society. Baudelaire's idea of the modern, articulated in his essay *The Salon of 1845*, was 'anything that attracts the eye of the crowd and the artists': it was no longer to be found in academies, galleries or drawing rooms, but among the teeming streets. As the cheap and mechanical trick of photography replaced the art of portraiture, the modern painter was obliged to be 'a genius who is ceaselessly in search of the new', and modern art a Romantic pursuit of 'intimacy, spirituality, colour, aspiration towards the infinite'.[78] The artificial paradises of hashish, and other drugs yet to be discovered, would be subsumed into the coming vogue for sensation.

By the end of the century Paris would present a spectacle unlike anything the world had seen before. Public boulevards, parks and squares merged with the commercial world of cafés, restaurants and nightclubs to create a grand public stage for promenade, leisure and entertainment, with an ever-changing cast of self-fashioning performers and voyeurs, each finding in the crowd their freedom to be an individual. Culture in all its forms spilled out onto the streets: arcades, shopfronts and kiosks plastered with vibrant, jarring posters were the new galleries. In 1881 Guy de Maupassant would describe 'the Mentality of the Boulevards' as an altered state of consciousness in its own right – a 'mad mental condition – quarrelsome, flippant, whirling and emptily sonorous' – that would subsume all the denizens of the modern city in its thrall.[79] As it did so, the Oriental myths that had framed the hashish experience would be succeeded by a new, distinctively modern playground of the senses.

CHAPTER SIX

EXTASIA, FANTASIA AND ILLUMINATI

The hashish vogue of mid-century Paris diffused across the globe in surprising ways. It would have been hard to predict, for example, that by the 1860s the largest importer of hashish to the United States would be a Black Rosicrucian sex magician.

Paschal Beverly Randolph grew up in New York with, in his own words, 'the great disadvantage of an unpopular complexion, and a very meagre education to back it'.[1] His father, who abandoned him, was a white man from Virginia; his mother, who died in the almshouse of Bellevue Hospital during the 1832 cholera outbreak when he was seven, was by his account a descendant of Madagascan royalty. Randolph was raised on the streets of Five Points, the heart of the slum district, which he escaped by becoming a cabin boy on ships trading with Cuba. He was invalided out of service and moved to Portland, Maine, where he worked as a dyer and a barber before settling in Utica and joining the spiritualist community that was booming in upstate New York after the Fox sisters and their table-rapping séances became a global sensation in 1848. He proved to be a captivating trance speaker, addressing audiences in the channelled voices of famous figures ranging from Zoroaster to Benjamin Franklin. By 1853 he was billing himself in the Utica city directory as 'Dr Paschal Beverly Randolph, clairvoyant physician and psychophrenologist', with a special expertise in treating marital problems, which he believed were at epidemic levels on account of a rigid patriarchal society that ignored women's needs and desires.[2]

22. Paschal Beverly Randolph: occultist, sex therapist, trance medium, prolific author and purveyor of hashish-based elixirs for clairvoyance and healing.

The following year Randolph moved back to New York City and introduced himself to the staff of the *Spiritual Telegraph*, America's first spiritualist journal, with a compelling trance performance in which he channelled the voice of Robespierre. He was now positioned at the centre of a network that supported medical consultants, writers and lecture tours, and he was invited to travel to Europe to attend the 1855 World Convention of Robert Owen's socialist and spiritualist disciples. He addressed a spiritualist circle in London's Charing Cross, where he appropriately channelled Sir Humphry Davy, and then on to Paris, where he was given introductions to mesmerists and trance mediums. This network merged with that of learned occultists including Alphonse Louis Constant who, writing as Eliphas Lévi, had just published the first volume of his *Dogme et rituel dela haute magie*, which would become the bible of nineteenth-century ceremonial magic.

Randolph was one of the first American visitors to occult Paris, and the exchange was a fruitful one. For the previous generation, the magical arts had been largely the pursuit of antiquarian scholars rather than those seeking direct personal contacts with spirits or supernatural powers. Randolph was voraciously self-educated in magical traditions, but his engagement with them was practical, embodied and theatrically performed, and he found among the French mesmerists a variety of tools and techniques unknown in the United States to assist his immersion in the oracular state of trance. One was the magic mirror, a convex glass with a dark surface in which scenes could be witnessed by clairvoyance: subjects would gaze into it while under trance, having previously magnetised it with mesmeric passes. Randolph added mirror magic to his repertoire and developed his own refinements, including smearing the surface with genital secretions for sex magic.

He was also introduced to what would become his defining magical aid, hashish. His initiator was a cabinet maker and furniture restorer named Louis-Alphonse Cahagnet, a Swedenborgian mystic who used mesmerism and somnambulism to explore the spirit realm. He had begun by eliciting messages from mediums but grew frustrated by hearing of angelic encounters and life after death only at second hand. He had considered the use of drugs, but tentative experiments with ether and the fumigants of ancient magical tradition such as belladonna and hemlock convinced him that they were too crude and toxic for the subtle energies of the spirit: 'narcotics carry problems into the nervous system, they trouble the soul in its vital functions'.[3] In 1849, however, his interest was piqued by the extravagant reports of the Club des Hashischins and the availability of hashish in new and convenient pharmacy preparations. He bought some, and arranged a séance with some friends:

> I swallowed three grams of the extract of hashish, dissolved in a cup of coffee. At the end of an hour to an hour and a quarter, I experienced the first effects of this *medicament of the soul.* I was thinking of the laughter that the hashish always occasions as its first symptom, and this thought of laughter prompted me to laugh wholly against my will . . . A moment of calmness succeeded this first attack, but it was not of long duration. I felt myself at once seized with a sort of universal intoxication . . . I then entered completely into the ecstatic state, my eyes being open, and having perfectly the consciousness of my existence and my condition.[4]

As Jacques-Joseph Moreau had described, the first effect of the drug was euphoria and general hilarity; but as the intoxication took hold, this was succeeded by the *état mixte*, in which, Cahagnet wrote, 'I then really commenced two perfectly distinct lives, the acts and tableaux of which succeeded each other without confusion, with order and regularity'.[5] Gradually the bombardment of fantasies eased, until with an effort he was able to focus on a persistent vision that became a revelation:

> I saw as it were an immense vortex of an immeasurable depth, having the form of an ellipse. One of the centres of this ellipse was white and wonderfully luminous; everything revolved around the centre . . . All was in motion around this focus . . . It seemed to me that this focus of light, this centre of universal attraction, was *God*; that he was the link of all beings and the reason of their existence.[6]

After this the drug's influence became once again unruly and chaotic, assailing him with 'grotesque visions, such as cascades of teeth, afterwards forming men's heads, and other pantagruelic fantasies', which Cahagnet involuntarily interrupted with 'incessant bursts of laughter, which changed by degrees into cries and howls having nothing agreeable in them for the listener'.[7] But the vision of divine light was precisely what he had been seeking for many years, and he became an apostle for hashish as a conduit to a parallel reality that rational thought had obscured. 'Let us hope', he wrote, 'that in a short time the words *madness*, *hallucination* and *imagination* will be erased from our scientific language to be replaced by those of *internal life* and *external life*.'[8] He began to hold hashish séances at his rooms in the Rue St Denis and recorded a series of transcendent visions, from which one of his subjects, a journalist named M. Mouttet, emerged with a conclusion worthy of William James: 'The spiritual world is *in the material world*; it is another mode of the soul's seeing – a STATE. There are only STATES. To each state belongs its own perception.'[9]

For his part, Paschal Beverly Randolph wrote of his discovery of hashish:

By its aid Alphonse Cahagnet, myself, and others, have been enabled to pass through eternal doors, forever closed to the embodied man save by this celestial key, and passing through them, in holy calm, to explore the ineffable and serene mysteries of the human soul, and attain unto a conviction of immortality.[10]

'Dream begins where the freedom to direct our thoughts ceases,' Moreau had written, 'a new life follows the other and replaces it.'[11] The limits of the will marked the boundary of the self, and the gateway to the world of dreams; but the effect of hashish, in large enough doses, was to subdue the will and give free rein to a second self, with heightened senses and imagination. Cahagnet and Randolph believed that magical practice could bring the world of dreams under the control of human agency. The intoxication of hashish was overwhelming and exhausting, but with a properly trained mind it could, temporarily at least, be directed towards contact and dialogue with beings and dimensions beyond the reach of everyday consciousness.

* * *

Over the next few years Randolph continued to shuttle between Europe and the United States, where he re-established his medical practice in Utica in partnership with his wife, Mary Jane, who had her own practice in Boston treating women's problems with remedies drawn from her claimed Native American heritage. (As with her husband, Mary Jane's ethnicity is disputed: both may have been concealing an Afro-American slave ancestry, a common strategy in the antebellum United States, where Black spiritualists and healers often reinvented themselves as indigenous medicine folk or Oriental fakirs.) In 1861 Randolph made a lecture tour of California before undertaking another journey to Europe and onward to the Orient, where he spent nearly a year in Turkey, Egypt and Syria. He immersed himself in the world of dervishes, healers and miracle workers, and reports filtered back that he had been initiated into 'many of the dark secrets of the Oriental magicians'.[12] He sought out suppliers of hashish and *dowam meskh* (his spelling of *dawamesc*), and claimed to have learned the secret of his favoured preparations from the Sultan's physician in Istanbul.

On his return, Randolph's experiences in the Middle East came to define his public persona, and his bespoke hashish elixirs became his most distinctive product. His travels echoed an episode in the biography of Christian Rosenkreuz, the legendary seventeenth-century founder of Rosicrucianism, whose missing years spent travelling in the Orient were said to have included an initiation into the ancient wisdom of the Magi. Randolph's writing took on a haughty exoticism, sprinkled with references to *Allah* and *kismet* and invocations to cryptic dignitaries such as 'Abu-id-Durr, Djundub of the Ansarieh':[13] a style reminiscent of his contemporary Sir Richard Burton, the British explorer whose translation of *The Arabian Nights* scandalised Victorian Britain with its combination of cranky erudition, floridly archaic prose and prurient focus on drugs, squalor and pederasty.

Back in the United States, Randolph found that hashish eaters' tales were enjoying a warm reception among the reading public. The poet and travel writer Bayard Taylor's 1855 account of his Middle Eastern travels, *The Lands of the Saracen*, went through twenty editions over the next ten years and established the genre of drug reportage to which Silas Burroughs would make his contribution four decades later. In his chapter 'The Visions of Hasheesh', Taylor recounted a night on a hotel roof in Damascus where he and a couple of fellow travellers each ate a large lump of hashish, a dose that Taylor later estimated was 'enough for six'. Soon enough,

> The Spirit (demon, shall I not rather say?) of Hasheesh had entire possession of me. I was cast upon the flood of his illusions, and drifted helplessly whithersoever they might choose to bear me. The thrills which ran through my nervous system became more rapid and fierce, accompanied with sensations that steeped my whole being in unutterable rapture. I was encompassed by a sea of light, through which played the pure, harmonious colors that are born of light.[14]

Taylor found himself transported to the base of the Great Pyramid of Cheops, which he ascended in his imagination to its summit, just as Silas Burroughs would. As he looked down over the palm groves, he noticed that the monument on which he stood 'was built, not of lime-stone but of huge, square plugs of Cavendish tobacco!' A paroxysm of hilarity gave way to a vision in which:

> I was moving over the Desert, not upon the rocking dromedary, but seated in a barque made of mother-of-pearl, and studded with jewels

of surpassing lustre. The sand was of grains of gold, and my keel slid through them without jar or sound. The air was radiant with excess of light, though no sun was to be seen. I inhaled the most delicious perfumes; and harmonies, such as Beethoven may have heard in dreams, but never wrote, floated around me. The atmosphere itself was light, odor, music; and each and all sublimated beyond anything the sober senses are capable of receiving.[15]

Yet this fantastical scene, seemingly elaborated from the magic carpet of *The Arabian Nights* but vividly embodied in all his senses, somehow coexisted with his waking consciousness:

The most remarkable feature of these illusions was, that at the time when I was most completely under their influence, I knew myself to be seated in the tower of Antonio's hotel in Damascus, knew that I had taken hasheesh, and that the strange, gorgeous and ludicrous fancies which possessed me, were the effect of it.[16]

By midnight, his senses overloaded and his nerves shredded, Taylor had passed from the 'Paradise of hasheesh' into 'its fiercest hell', a torment from which he only recovered after a sleep that lasted thirty hours. The episode concluded with a brief sermon on 'the awful peril of tampering' with 'the majesty of human reason and of human will'. This was the message conveyed by Thomas De Quincey, with his famous insistence that the pleasures of drugs were inseparable from their exquisite pains, and reinforced by Charles Baudelaire's declaration that this was a 'forbidden game', in which base instincts would always triumph over noble motives. As drug reportage evolved into a recognisable literary genre, this proved an enduringly popular trope. It allowed publishers to sell immorality while absolving them of the charge of promoting it, and authors to encourage the belief that the paradises of opium or hashish were real and attainable, but only to the elite few prepared to plumb the depths of hell. With the first home-grown American tale in this genre, the rules were already established.

In 1857 a fresh literary sensation emerged from upstate New York as an American successor to De Quincey and Baudelaire. *The Hasheesh Eater, or Scenes from the Life of a Pythagorean* was the eye-catching debut of a twenty-one-year-old prodigy, Fitz Hugh Ludlow, a recent graduate of Union College, Schenectady, in which he described how his experiences with hashish had come to constitute a parallel existence, played out in the now familiar landscape of *The Arabian Nights*:

> There Oriental gardens waited to receive me. From fountain to fountain I danced in graceful mazes with inimitable houris, whose foreheads were bound with fillets of jasmine. I pelted with figs the rare exotic birds, whose gold and crimson wings went flashing from branch to branch, or wheedled them to me with Arabic phrases of endearment. Through avenues of palm I walked arm-in-arm with Hafiz, and heard the hours flow singing through the channels of his matchless poetry. In gay kiosks I quaffed my sherbet, and in the luxury of lawlessness kissed away by drops that other juice which is contraband unto the faithful.[17]

In waking life, Ludlow had travelled no further east than Brooklyn. His initiation into hashish came as a teenager in Poughkeepsie, where his father was a Presbyterian minister. He befriended the town apothecary and spent hours rummaging in his stores, surreptitiously sampling his chloroform and ether. When a shipment of cannabis extract arrived, Ludlow read up on it in James Johnston's recently published textbook *The Chemistry of Common Life*, which included a helpful section entitled 'The Narcotics We Indulge In'. The chapter on Indian hemp was a thorough survey that discussed the legend of the Assassins, explained the difference between preparations such as *bhang*, *charras*, hashish and *dawamesc*, and quoted extensively from the work of O'Shaughnessy, Moreau's self-experiments and the discoveries of the French pharmacists.

'Upon Europeans', Johnston concluded, 'its effects have been found to be considerably less in degree than upon Orientals,' an opinion that Ludlow was quick to put to the test.[18] His first dose was too cautious to have any effect, but his second, of 2 grams, plunged him into chaotic hallucinations. He stumbled to the local doctor's surgery, hissing that he had taken an overdose of hashish and was about to die; the doctor replied 'Bah!' and offered him a sedative. Tormented by visions of malign gnomes, he retreated to his bed where the turmoil eventually gave way to blissful *kif*, and the realisation that his life had been set on a new course:

> In the presence of that first sublime revelation of the soul's own time, and her capacity for an infinite life, I stood trembling with breathless awe. Till I die, that moment of unveiling will stand in clear relief from all the rest of my existence.[19]

The following year, leafing through *Putnam's Magazine* in a bookshop, he stumbled with amazement on Bayard Taylor's 'Vision of Hasheesh', and pronounced that 'this man has been in my own soul!'[20] His own book took shape as an arabesque account of his visions, heavenly and hellish, interspersed in De Quinceyan style with episodes from his life and philosophical excursions in which he claimed that hashish gave access to the Pythagorean and Platonic world of ideal forms, and that Pythagoras must himself have been a hashish initiate to have grasped the numerical order underlying the cosmos.

Ludlow wrote that 'I was suckled at the breast of Transcendentalism';[21] Walt Whitman would become a mentor to him, and *The Hasheesh Eater* is the most substantial exploration of the overlap between the metaphysics of that school and the drug experience that it typically denounced. By the end of the book, however, Ludlow had arrived at a position closer to Emerson's: 'The motives for the hasheesh-indulgence were of the most exalted ideal nature', but, he concluded, in reality it was 'an accursed drug, and the soul at last pays a most bitter price for its ecstasies'.[22] It carried him to literary fame, and he became a regular in New York's small bohemian scene, hobnobbing at Pfaff's bar on Broadway with Walt Whitman and Mark Twain; but, like his model De Quincey's masterpiece, it turned out to be the prelude to a career of often scrappy short fiction and journalism, and a life of precarity, debt and increasing dependence on the opium to which he resorted to soothe his tuberculosis, insomnia and neurasthenia. His gambit of launching himself as 'The Hasheesh Eater' became, like De Quincey's, his inescapable literary brand, defining him until his early death at the age of thirty-four.

*   *   *

In the United States to which Paschal Beverly Randolph returned in 1862, hashish had two distinct identities: a source of Oriental visions of sublime beauty and terror, and simultaneously – usually under the name of Indian hemp – a pharmacy product, one among many preparations recommended for analgesia and anti-spasmodic relief. Randolph fused these identities, and borrowed the literary stylings of Taylor and Ludlow, to create and market a range of hashish-based elixirs and tinctures that, like the magic mirrors he sold alongside them, came with claims of unmatched quality and potency, and promises of clairvoyant powers. He promoted them through classified advertisements in spiritualist magazines such as the *Spiritual Telegraph* and the *Banner of Light*, whose readers were informed on his return from Europe that:

> In reply to numerous correspondents, let me say that nearly all the Hashish I brought with me from Europe (and none other is fit to use) is exhausted. The balance I will sell at four dollars a bottle, with full directions how to secure the celestial, and avoid the ill fantasia.[23]

The precise contents of Randolph's product range are hard to reconstruct, concealed as they were behind magical claims, florid epithets and veils of commercial secrecy. His elixirs included pure *dowam meskh* alongside 'protozone' and 'protogene', which seem to have been hashish-based tonics for nerves and sexual health, and a 'universal solvent, or celestial Alkahest', which may have been a branded variant of the *dowam meskh* intended for clairvoyance.[24] He may also have included hashish in the 'triple strength' aphrodisiac potions he sold as 'Phymylle' and 'Amylle': 'liquids more potent than essential wine . . . the fluid draft of power itself'.[25] Some of his elixirs may also have contained belladonna, henbane or opium, either for consumption or for ceremonial purposes, such as smearing on magic mirrors. He stressed that the hashish preparations available in pharmacies were 'ordinary extracts, possessing a little medicinal power, and a great deal of anodyne, soporific, brain-maddening force', and were not to be compared or confused with elixirs 'prepared by adepts, whose art has been handed down to them from centuries past', for the 'luxurious lordlings of the Orient'.[26]

Randolph's most explicit descriptions are found in a short pamphlet titled *Hashish: Its Uses, Abuses and Dangers, its Extasia and Fantasia, and Illuminati*, privately printed and later included in his 1867 *Guide to Clairvoyance*. In it he advises those preparing to take it to be ready for 'imaginations turned to realities of the queerest, strangest, weirdest and perhaps terrific kind'. Those taking it should have six lemons and 2 ounces of citric acid to hand as antidotes in case of emergency. If hashish was taken thoughtlessly, 'the Extasia will rush through the nerves' and leave the subject 'convulsed with laughter, terror, horror, fear of death'. With the correct mental preparation, however, Extasia will give way to Fantasia, a state in which:

Tables talk, ordinary rooms become magnificent palaces, and the most common things and objects are totally transformed . . . Beyond all doubt, the 'Arabian Nights' romances were the results of so many doses of hashish, penned as the visions occurred . . . There is no doubt that Confucius, Pythagoras and his disciples, the Alchemists, Hermetists, Illuminati, and mystic brethren of all ages used it to exalt them while making their researches for the Philosopher's Stone.[27]

Randolph swirled his spiritualist, occult and literary influences into intoxicating prose, but behind it can be discerned the same practical considerations that Cahagnet had first reckoned with. Hashish hurls its subject into a vivid and chaotic inner world, but one over which the conscious mind is still able exert control and will. Harnessing this state – in Randolph's terms, conquering the Extasia and channelling the Fantasia – is a mental and spiritual practice that demands skill and courage, comparable to breaking a headstrong horse. Maintaining control is exhausting, and impossible to sustain throughout the long hours of hallucination; the goal is to achieve it for long enough to receive a glimpse of illumination. Randolph conceived clairvoyance, or 'sympathy', as an altered state of consciousness, 'a magnificent sweep of intellect that leaps the world's barriers', the highest level of human functioning but one accessible to all.[28]

Hashish was only one of the sources of spiritual wisdom that Randolph offered his customers. He travelled and lectured energetically, speaking in his own voice rather than the trance personalities of his early career. He wrote and published continuously, recycling his lectures and plagiarising the work of others, using multiple pseudonyms, leaving a bibliographic trail that has still not been fully itemised.[29] He continued his medical practice, sometimes undertaking consultations under trance. He distanced himself from spiritualism and styled himself as a Rosicrucian, settling in Boston and billing his consulting offices as 'the Rosicrucian Rooms'.[30] He involved himself in politics and social issues, writing in his book *Pre-Adamite Man* (1863) that the Negro race was the purest strain of humanity and destined for future ascendancy. He was championed by some of Boston's abolitionists and corresponded with local mystics and orators including Benjamin Paul Blood, whose anaesthetic revelation was unfolding nearby and at the same time.

Despite his eloquent salesmanship, Randolph was ambiguous in his advocacy for hashish. In his later career he turned against it – 'I do not approve of the use of hashish, for extasia, fantasia, or clairvoyance, any more' – and even claimed that he had only taken it four times in his life: 'twice on purpose, and twice accidentally, many years ago. I have not used it since, not that I fear its power, but because I need it not.'[31] The final years of his life were troubled by inebriety, injury and legal troubles, and eventually ended by suicide, but the role of hashish in them is hard to interpret. It may simply be that, following Taylor and Ludlow, he judged it politic to offset his advocacy for the drug with stern warnings of its consequences. It may be that he found the trifecta of sex, drugs and his ethnicity conferred too heavy a stigma; many Black religious and community leaders kept him at arm's length on account of his licentious reputation. It may be that his personal pursuit of the hashish fantasia offered diminishing returns. In 1870 he sold his medical practice, and the elixirs disappeared from view.

*   *   *

The decline of traditional religious observance was among the most conspicuous social changes of the late nineteenth century, but one of its less examined consequences was that drug experiences took on dimensions of the sacred that had previously been part of its domain. The mystical revelations produced by nitrous oxide, ether and chloroform were opened up to scientific experiment, and spiritual interpretations of them were obliged to compete with modern notions of consciousness, the subliminal self or the pluralistic universe. Conversely, the experiential and embodied form of magic popularised by Paschal Beverly Randolph cast intoxicants such as hashish as magical keys to other worlds. With the retreat of prayer, the West's traditional technique for entering an altered state of consciousness conducive to meditation and receiving spiritual grace, new practices emerged to expand subjectivity, enhance the imagination and open the mind to immaterial influences.

In 1854, shortly before Randolph's arrival in Paris, Eliphas Lévi conducted a ritual invocation of the ghost of the ancient Greek magus Apollonius of Tyana that became one of the founding events, or legends, of the nineteenth century's occult revival. Lévi refers to 'horrible' substances that confer the power of 'dreaming while awake', and mentions 'aconite, belladonna and poisonous mushrooms'.[32] His description of the rite includes two braziers that he filled with the 'required and prepared substances' while dressed in 'robes like that of our Catholic priests', after which:

> The white smoke slowly rose above the marble altar. I seemed to feel the earth tremble; my ears were ringing and my heart beat strongly. I put several branches and perfume in the braziers, and as the flames rose again, I distinctly saw, in front of the altar, the larger-than-life figure of a man, which then dissipated and faded away . . . I closed my eyes and called out to Apollonius three times; and when I opened them again, a man stood before me . . .[33]

Lévi was cryptic about the precise plants and essences involved in his conjuration, and elsewhere he writes that the use of narcotics in ceremonial magic is misguided and potentially dangerous.[34] He may have been referring to incenses and perfumes rather than narcotics or hallucinogens; if 'horrible' intoxicants were involved, they could have been incorporated into the ritual without actually being ingested. Nonetheless, the magical orders that emerged in the late nineteenth century frequently recalled the vision-giving potions and fumigants of ancient priestesses, Persian sorcerers and medieval witches. By this time, fumigant preparations of datura and belladonna were not the exclusive preserve of magical rites, but widely available as pharmacy products for treating asthma.[35] As spiritual practices devolved into domestic settings, these profane substances suggested themselves for sacred uses.

The idea that intoxicating plants might be sacred or magical found some support in the writings of Madame Blavatsky, the founder of Theosophy. She taught that some plants had 'mystic properties in the most wonderful degree', for example the sacred *soma* potion described in the *Rig Veda* or the sacrament of the Eleusinian mysteries in classical

Greece, but that their secrets had been 'lost to European science'.[36] She had a distant but abrasive relationship with Paschal Beverly Randolph: they may have met in person in the USA, but much of their communication appears to have been telepathic (he referred to her as 'the Old Lady', she called him 'the N-----'.)[37] Most of Blavatsky's followers took her to be opposed to the use of drugs, but some, particularly those with a close connection to the aesthetic and decadent currents of the *fin-de-siècle*, regarded her assertion that their magical secrets had been lost as a tacit charter to experiment and rediscover the 'poison path'.

The occult revival of this era was inspired by lost or forgotten religious knowledge such as John Dee's Enochian keys, the Hebrew kaballah and the Theosophy of Jacob Boehme, but it also incorporated currents of progressive, modern and scientific thinking. Many of its initiates were interested in the latest researches into double consciousness and the subliminal self, which opened up novel possibilities for introspection, harnessing irrational forces and cultivating psychic powers. Eastern spiritual disciplines such as Rāja and hatha yoga pointed the way to new techniques for achieving higher consciousness and accessing the ethereal or astral plane. Madame Blavatsky's Theosophy drew on Darwinian evolution for its cosmic narrative, and the occult journals of the 1890s were among the first to offer translations of and commentaries on Friedrich Nietzsche's writing on the coming age of the Superman. Drugs such as opium, belladonna and particularly hashish recurred through these disparate sources: they were simultaneously ancient and modern, Eastern and Western, and agents of still untapped human possibility. A survey of hashish in the spiritualist journal *Light* in 1893 concluded that 'The faculty most influenced by this narcotic is the imagination'. Under its influence, the everyday world could be 'idealised into the most beautiful and fantastic forms'.[38]

Officially, the magical orders that emerged in the 1880s were united in disavowing the use of drugs.[39] Most of their recruits came from the respectable worlds of Freemasonry, Rosicrucianism and Theosophy, where they had been taught that drugs were unnatural and deleterious, sapping and poisoning the natural exercise of the will. The Hermetic Order of the Golden Dawn, the best-remembered magical society of the era, was a case in point. Some of its famous members, such as the Irish poet W.B. Yeats, used hashish and other drugs in their spiritual practices, but they did so outside its immediate circle.

MacGregor Mathers, co-founder and leader of the Golden Dawn, had previously been raised to the rank of master mason in his Hampshire lodge and was ambitious for his Order to recruit distinguished literary and cultural figures into its membership. Yeats was one of his early successes: he had become a student of Theosophy in Dublin at the age of twenty, but left the society in 1890 rather than take a compulsory pledge of loyalty to Madame Blavatsky. (Mathers would demand similar pledges from some Golden Dawn members, but Yeats was omitted from this list.) Yeats found Mathers 'imperfect' in his tastes and cranky in his scholarship – to the amusement of Yeats's sophisticated friends, he believed the poems of Ossian to be ancient Gaelic long after they had been exposed as an eighteenth-century fraud – but the ceremonial magic he practised was a rich source of symbolism and insight that Yeats credited as a major influence on both his magic and his writing.[40]

Mathers, like most Freemasons, regarded the use of drugs as low-grade and profane: in the words of Aleister Crowley's secretary Israel Regardie, he 'frowned upon all such methods, preferring the classical secret techniques of mind and spiritual training'.[41] But Yeats inhabited a set of overlapping social circles in which hashish and other drugs were imbued with a more exalted sense of possibility. As well as the magicians of the Golden Dawn, he was connected to the experimenters of the Society for Psychical Research, of which he was an associate member, and was a founder member of the Rhymers' Club, a London grouping of bohemian poets who met in the Olde Cheshire Cheese pub on Fleet Street.

It was through the Rhymers that Yeats was introduced to hashish in 1890, around the time he joined the Golden Dawn. His friend and fellow poet, Arthur Symons, was the drug's most conspicuous devotee in

London's small and tightly knit bohemian and decadent coterie. Symons was a devotee of Charles Baudelaire and the French symbolist poets and, despite or because of his stern religious upbringing by Cornish Methodists, he was the London literary figure who flirted most conspicuously with the transgressive currents of *fin-de-siècle* Paris. The ethos of the Rhymers' Club was fiercely antagonistic towards the bland 'mass cult' of commercial publishing, and saw itself as a refuge for writers driven by a pure vision of their art.[42] In the upper room of the Olde Cheshire Cheese they read their poetry to one another and ignored the tawdry modern world; afterwards, Symons would invite members back to his rooms in nearby Fountain Court, in the Temple district, where the chambers of the London legal profession clustered beside the Royal Courts of Justice. He was known to retire there with the decadent and hard-drugging poet Ernest Dowson, with whom he indulged in 'haschisch – that slow intoxication, that elaborate experiment in visionary sensations'.[43] On one occasion, scandalously, he invited 'some of his lady friends from the ballet' to take some with tea.[44] His description of its effects expanded on Baudelaire's, entwining the magical, the aesthetic and the transcendental:

> After indulging in it, one sits, as in a theatre, seeing a drama acted on a stage. We see it all with eyes that – during these ecstatic hallucinations – behold an endless drama of dreams; that perceive the subtlest impressions, fairy pageants, ghostlike unrealities: eyes, in short, that envisage the borders of the infinite . . . The instant becomes eternity; though the hallucination is sudden, perfect and fatal. One feels an excessive thirst, a physical restlessness, a nervous apprehension, which at last subsides into, that strange state which the Orientals call *Kief*.[45]

Symons saw London through Baudelaire's modern eyes, as a flux of dramatic scenes and images, a kaleidoscope of life. He would walk at night from Fountain Court into Soho and Covent Garden, where he obsessively toured the music halls and dance revues, and then 'wander into absolutely unknown regions', the slums and rookeries of St Giles and beyond.[46] His 1895 collection of poems, *London Life*, was an arabesque of dreamlike, fragmented impressions under fog and gaslight, juxtaposing tawdry beauty with ugliness and vice; it was his most scandalously decadent work, turned down by respectable publishers including John Lane and William Heinemann. But Paris, he was obliged to concede after his first visit in 1889, was 'the real thing'.[47] The Olde Cheshire Cheese, with its bare sanded floor, draught ale, meat pies and pipe-smoking in its upstairs rooms, was adequate to host a handful of poets self-exiled to the distant fringes of the city's cultural life; but it was a provincial club compared to the nightlife of Montmartre, where Symons rubbed shoulders with the poet Paul Verlaine in his corner at the Café François Premier and was invited to the glittering salons at Stéphane Mallarmé's apartment on the Rue de Rome. It was to Paris that Symons took Yeats in 1890 for his first hashish experience.

For British visitors, even those from London's literary avant-garde, Paris was a destination of high adventure. It was a palace of the senses, the epicentre of artistic taste, fashion and high culture; it was a society of radical tolerance by British standards, where behaviour that would lead to ostracism or even prison in London was barely noticed. Candid portrayals of sex and drugs, banned at home, were celebrated in its art, poetry and confessional literature, and the nightlife – the gore-drenched Grand Guignol theatre, and the brazenly seductive cabaret of the Moulin Rouge and the Folies Bergère – pandered to pure sensation. Living was far cheaper, making it an attractive resort for those exiled from Britain by scandal or poverty.

Symons introduced Yeats to Paul Verlaine, and conducted him to the backstreets of the Latin Quarter, where they met with a group of Martinist mystics that included 'a boisterous young poet' who gave him a pellet of hashish before they went out to dinner, and a second one on his return.[48] Yeats had a brief flash of ecstatic transport before being brought down to earth by the poet's ostentatious prattling. He closed his eyes, awaiting visions that never came, though when he opened them he found the colours in the room strangely delightful, and 'it struck me that I was seeing like a painter'. He was annoyed by another young hashishin who 'ran towards me with a piece of paper on which he had drawn a circle with a dot in it, and pointing at it with his finger he cried out "God!, God!"'. The fastidious Yeats felt awkward in this vulgar and garrulous company, and as a consequence: 'I never forgot myself, never really rose above myself for more than a moment, and was even able to feel the absurdity of that gaiety, an Herr Nordau among the men of genius but one that was abashed at his own sobriety.'[49]

In 1894 Yeats took hashish in Paris again with Maud Gonne, the object of an unrequited passion that would last for decades. They had first met the previous year when she visited his family home in the west London arts-and-crafts suburb of Bedford Park to canvass his father's support for Irish independence, the revolutionary cause to which she devoted her life. Yeats was thunderstruck by her statuesque beauty – 'her complexion was luminous, like that of apple blossom through which the light falls' – and captivated by their shared passion for Celtic mysticism and magic.[50] In 1896, during a visit to Galway and the Aran islands with Symons, his obsession with Gonne reached a frightening pitch: he had an urge to plunge his hand into a fire to prove his devotion to her, and wrote later: 'I wonder at moments if I was not really mad.' For nine successive nights at Tulira Castle in Galway, he solemnly invoked the forces of the moon in an empty chamber, and received a hypnagogic vision of 'a marvellous naked woman shooting an arrow at a star'.[51] According to his biographer Norman Jeffares, his infatuation with Gonne, together with the influence of MacGregor Mathers and the Golden Dawn, drew him into 'the unreal maze of magical speculation' and left him 'on the verge of dissipation, taking hashish, and within the edge of that state, where he could easily have taken permanently to drink'.[52]

Yeats introduced Gonne to the ceremonies of the Golden Dawn; she was eager to learn its secrets, but unimpressed by its membership. As an independent 'New Woman' and a staunch anti-colonialist, she regarded the Order as the epitome of 'British middle-class dullness', overlaid with an 'English love of play-acting and grand-sounding

23. *Maud Gonne used hashish for meditation, astral travel and telepathy experiments with W.B. Yeats.*

titles'.[53] Nonetheless she absorbed some of their practices into her own magical experiments in astral travel, Celtic second sight and, with Yeats, telepathy. Gonne was an occasional user of various drugs, and publicly endorsed Vin Mariani's coca wine as part of her political campaign: 'Your coca wine, by fortifying my voice, will allow me to make my country better-known.'[54] She experimented with chloroform and found that it could help 'get her out of her body', but she worried about becoming addicted to it, and discovered that she could achieve similar results simply by imagining and visualising its effects.

On occasion Gonne also used hashish, which had 'convinced me of the possibility of being able to leave the body and see people and things at a distance, and to travel as quick as thought'.[55] She found it effective for controlled visualisation or self-hypnosis, and under its influence meditated on the *tattva*, a Sanskrit term for element or essence that provided a set of keys for travelling in the astral realms. The system had been elaborated by the Theosophists into a series of 'thought-forms', and was one of the techniques Yeats learned from Mathers for inducing clair-voyance: he made himself a set of cards for its five elements of earth (a yellow square), air (a blue circle), fire (a red triangle), water (a silver crescent) and spirit (an indigo egg).[56] Gonne, on hashish, visualised an indigo egg with a yellow square in the centre – her body surrounded by its spirit – merging with a red triangle. She drifted in reverie through the essences, symbols, shapes, elements and colours; as the material world fell away, she created a gateway into the astral dimension. Here, elemental spirits might be encountered; they were to be greeted respectfully and with unwavering self-control. The following morning fresh ideas and inspiration would arrive.

One night in Paris, after a large dose of hashish, 'that strange Indian drug', she:

found my legs paralysed; I could not walk and my heart was beating queerly, but my brain was clear . . . I managed to scribble on a writing pad I had on the table beside my bed that I had taken haschish and that, if I did not wake, no-one was to be blamed, and then I lay still and waited. I saw a tall shadow standing at the foot of my bed and it said, or more exactly, the thought drifted through my mind: 'You can now go out of your body and go anywhere you like but you must always keep the thought of your body as a thread by which to return. If you lose that, you may not be able to return.'[57]

She formed a wish to see her sister in Ireland, and was instantly transported there, but as she wandered around the dark and silent house it morphed into an unfamiliar and disturbing place, and she remembered the injunction to return to her body; 'then, with the sensation of falling from a height, I was really lying in my bed, conscious of my heart pounding queerly'.[58]

Gonne's distaste for the Golden Dawn grew after they held one of their meetings in a Masonic hall, and she quizzed MacGregor Mathers on the Order's relationship to the Craft. He confirmed that some of their passwords were shared with its higher grades, at which point she resigned her membership on the grounds that 'Free Masonry as we Irish know it is a British institution and has always been used politically to support the British Empire'.[59] Gonne was a conspicuous example of the New Woman, liberated and opinionated, yet her semi-public use of drugs was exceptional even in these circles. Independent women were typically portrayed in popular culture as unstable and prone to hysteria, madness and suicide, and most were careful to project an image of wholesome health; in French art and literature in particular, portrayals of young women alluringly corrupted by morphine, ether or cocaine had evolved into a subgenre that merged with pornography. Gonne was inured to public opprobrium, and she persisted as an uncompromising campaigner for peace and Irish independence (and, after being jailed several times, for the reform of women's prisons) until her death in 1953 at the age of eighty-six.

Maud Gonne's disapproval was a factor in Yeats's estrangement from Mathers and the Golden Dawn, from which he distanced himself after

1899. In 1897 he took hashish in Paris again with Arthur Symons, and discussed Symons's long-gestating book on the decadent movement in literature, for which he had written some essays while working with Yeats, Aubrey Beardsley, Havelock Ellis and others as editor of the short-lived and controversial *Savoy* magazine. Yeats disliked the term 'decadence', with its overtones of contrivance for the jaded palate and, after 1895, the taint of Oscar Wilde's trial. He suggested 'symbolism', a term more commonly applied to painting, but one that resonated with the imaginal world from which Yeats's poetry and magic both drew. In his dramas, he used dance, drums and incantatory rhythmic verse to create 'the moment when we are both asleep and awake, which is the one moment of creation';[60] he took hashish to create the same ecstatic moment in his magical practice.

The Symbolist Movement in Literature, published in 1899 with a dedication to Yeats, was Symons's most enduring contribution to literary criticism. It claimed symbolism both as a rediscovery of ancient magic – 'Symbolism began with the first words uttered by the first man, as he named every living thing' – and as a modern form of expression for 'an unseen reality apprehended by the consciousness'.[61] Symons's symbolists, who included the *hashishins* Gautier, de Nerval, Baudelaire, Verlaine and Rimbaud, occupied a new territory between artist and magus, in which literature 'becomes itself a kind of religion, with all the duties and responsibilities of the sacred ritual'.[62]

* * *

It was in this receptive milieu that the first of what would later be called the major psychedelics arrived in Europe.[63] On Good Friday of 1897, while Yeats and Symons were in Paris, the art critic Havelock Ellis was staying, as he often did, in Symons's rooms at Fountain Court. Ellis, alone on a religious holiday, 'judged the occasion a fitting one for a personal experiment'.[64] He had recently read an account by the United States' leading neurologist, Silas Weir Mitchell, of his experience with peyote (also known as mescal), a Mexican cactus with vision-producing properties, and had succeeded in ordering some dried samples from the London pharmacists Potter & Clarke, best known for Potter's Asthma Remedy, a fumigant remedy made from powdered datura leaves. He made a decoction from what he understood to be a full dose, three dried cactus heads, and drank it slowly throughout the quiet afternoon.

Ellis was a physician by training as well as an aesthete, and his first report on his peyote experiment was a medical account for *The Lancet* that concentrated on its physiological effects. His subsequent, much fuller essay for the progressive literary quarterly the *Contemporary Review*, framed the novel psychedelic experience in terms closer to art criticism. Its title, 'A New Artificial Paradise', placed it in Baudelaire's lineage (and, half a century later, Aldous Huxley would draw inspiration from Ellis's aesthetic meditations for his seminal psychedelic essay *The Doors of Perception*.)[65] Ellis began with a careful ethnography of the cactus's indigenous use in tribal Mexico, noting that the Tarahumari people treated it as a god and celebrated it with 'a fantastic and picturesque dance'.[66] He experienced it, however, as a procession of 'what might be called living arabesques': 'a saturnalia of the specific senses and, above all, an orgy of vision', in which he inhabited a world of dazzling colours, shapes and textures:

> The visions never resembled familiar objects; they were extremely definite, but yet always novel; they were constantly approaching, yet constantly eluding, the semblance of known things. I would see thick glorious fields of jewels, solitary or clustered, sometimes brilliant and sparkling, sometimes with a rich dull glow. Then they would spring up into flower-like shapes beneath my gaze, and then seem to turn into gorgeous butterfly forms or endless folds of glistening, iridescent, fibrous wings of wonderful insects . . .[67]

Remembering that its indigenous adepts 'keep a fire burning brightly through their mescal rites', he turned on the gas lighting. As the shadows flared, pulsed and took on a life of their own, Ellis was 'reminded of the paintings of Claude Monet, and as I gazed at the scene it occurred to me that mescal perhaps produces exactly the same conditions of visual hyperaesthesia, or rather exhaustion, as may be produced on the artist by the influence of prolonged visual attention'. He drew extensively on his vast store of artistic references, comparing the visions to 'exquisite porcelain objects', 'the Maori style of architecture' or the delicate effects 'of lace carved in wood, which we associate with the *moucrabieh* work of Cairo'.[68] Recollecting the experience in tranquillity, he decided that 'if ever it should chance that the drinking of mescal becomes a habit, the favourite poet of the mescal drinker will certainly be Wordsworth', who described perfectly 'the halo of beauty which it casts around the simplest and commonest things'. Indeed, 'many of his most memorable poems and phrases cannot – one is almost tempted to say – be appreciated in their full significance by one who has never been under the influence of mescal'.[69]

Ellis's description of 'visual hyperaesthesia' was equally a hallmark of the *fin-de-siècle* culture that surrounded him. It reached an apogee in 1895 with August and Louis Lumière's screening of the first motion pictures in Paris, but cinema was only the latest iteration of an obsession with optical patterns and illusions that had built throughout the century. Since the 1820s, cheap devices such as the thaumotrope – a disc with images on both sides that seemed to merge as it spun – had been a popular catchpenny in markets and fairgrounds; the zoetrope, a mechanised version of the same illusion that transformed a sequence of images on its rotating drum into a moving loop, was conceived in the 1830s and mass produced as a device for home entertainment in 1866. The stereoscope, invented in 1838 by the British physicist Charles Wheatstone as a tool for laboratory research, was rapidly absorbed into mass entertainment: after Queen Victoria admired one at the Great Exhibition of 1851, stereo viewing devices sold in their hundreds of thousands, creating a vast popular market for images from astronomy to pornography.

From the beginning, these devices had been both scientific instruments and popular entertainments. Thaumatropes and zoetropes were marketed as 'philosophical toys', devices that simultaneously delighted the eye and boggled the brain, and in the era that followed Johann Purkinje's exploration of entoptic images they were regularly co-opted into the laboratory experiments of perceptual psychologists. All, in different ways, showed that moving images with no objective existence in the external world could be created in the mind. In 1824 Peter Mark Roget, of *Thesaurus* fame, who in his youth had participated in Humphry Davy's nitrous oxide trials, demonstrated by experiment the principle of 'persistence of vision' that underlay them. Charles Wheatstone used the stereoscope to demonstrate that the 3-D image it constructed had no real existence, but was fashioned by the brain from the differential between two flat photographs. David Brewster, the inventor of the kaleidoscope, saw it as an instrument for producing symmetry and beauty on a scale no human could match:

> There are few machines, indeed, which rise higher above the operations of human skill. It will create in an hour, what a thousand artists could not invent in the course of a year; and while it works with such unexampled rapidity, it works also with a corresponding beauty and precision.[70]

Baudelaire had described the *flâneur* or dandy, taking in the ever-shifting configurations of modern city life, as 'a kaleidoscope gifted with consciousness', simultaneously observing and creating 'the multiplicity of life itself and the flickering grace of all its elements'.[71] Under the influence of peyote, the human mind became a machine that generated illusions of equally prodigious rapidity and beauty, but drawn from a source deeper and more mysterious than the signals that shuttled between eye, optic nerve and brain.

Ellis was inspired to take peyote again, and was particularly eager to explore how its visions interacted with different forms of art. On this occasion he asked a friend to play the piano for him:

The music stimulated the visions and added greatly to my enjoyment of them . . . This was particularly the case with Schumann's music, for example with his *Waldscenen* and *Kinderscenen*; thus 'The Prophet Bird' called up vividly a sense of atmosphere and of brilliant feathery bird-like forms passing to and fro; 'A Flower Piece' provoked constant and persistent images of vegetation; while 'Scheherezade' produced an effect of floating white raiment, covered by glittering spangles and jewels.[72]

'Coloured hearing' had been described in classical Greece and studied by German psychologists since the early nineteenth century; but the scientific term 'synaesthesia' was freshly minted, by the American psychologist Mary Whiton Calkins in 1893. At the moment of Ellis's experiment, the phenomenon was of intense interest to scientists, artists and occultists alike. The first systematic survey of letter–colour associations had been undertaken by the German psychologist Gustav Fechner in 1876; Eugen Bleuler, who would later introduce the term 'schizophrenia' into psychiatry, labelled colours triggered by music as 'sound photisms', and the Swiss psychiatrist Théodore Flournoy had recently produced the first detailed case histories of synaesthetes and the meanings they assigned to their sensory peculiarities, or talents.

Thanks to Charles Baudelaire's famous account, it was already recognised that hashish could prompt episodes of synaesthesia, and that these were associated with ecstatic states and creative inspiration. Arthur Symons echoed Baudelaire with his description of the drug as 'a magician who turns sounds into colours and colours into sounds',[73] and in 1899 he evoked the rapturous state of sensory overload in his poem 'The Opium Eater':

I am engulfed, and drown deliciously
Soft music like a perfume, and sweet light
Golden with audible odours exquisite . . .[74]

Synaesthesia was taken as evidence, both scientific and artistic, that hashish – and now peyote – had the power to create sensory experiences that were unattainable in normal life to all but the exceptional few. As to its cause, psychologists offered competing theories. Some believed it arose from early childhood associations, others from abnormal brain structures, others still from a hereditary reversion to the evolutionary stage where animals had only a single sense organ.[75] This last suggestion was seized on by Max Nordau, the social critic whose book *Degeneration* (1892) positioned him as the leading scourge of aestheticism and decadent *fin-de-siècle* culture, who denounced synaesthesia as 'evidence of diseased and debilitated brain activity'. Raising it to 'the rank of a principle of art', he wrote, was 'to designate as progress the return from the consciousness of man to that of an oyster'.[76]

Occultists and spiritualists saw synaesthesia as evidence not of degeneration but of a higher form of consciousness, perhaps even the arrival of a new stage in human evolution. The 'thought-forms' of the Theosophists had colours that attuned their shapes with particular spiritual values: a book on the subject written by Annie Besant and Charles Leadbeater in 1901 systematised the colour meanings of musical notes and aligned them with the coloured shapes of the *tattva* essences. Artists, notably Wassily Kandinsky, believed colours had spiritual resonances and took inspiration from the brightly coloured abstractions of Theosophy's thought-forms. The composer and pianist Alexander Scriabin went as far as inventing the *clavier à lumières*, a coloured keyboard that mapped tones to notes, based on pairings in Madame Blavatsky's *Secret Doctrine*, which allowed visions to be transposed into music.

As well as musicians and artists, Ellis was naturally interested in Yeats and Symons's responses to peyote, and both joined him for sessions at Fountain Court, which he reported in lightly anonymised form. Yeats – a poet 'much interested in mystical matters, an excellent subject for visions' – found the cactus physically exhausting: 'he much prefers hasheesh'. Symons, however, was enraptured: 'I have never seen such a succession of absolutely pictorial visions with such precision.' When he played the piano with eyes closed, he 'got waves and lines of pure colour', but the most memorable moment came when he took a stroll down to the river Thames: 'Late in the evening I went out on the Embankment, and was absolutely fascinated by an advertisement of "Bovril", which went and came in letters of light on the other side of the river; I cannot tell you the intense pleasure this moving light gave me, and how dazzling it seemed to me.'[77]

In March 1897 Yeats passed on a sample of the new 'dream drug' to Maud Gonne.[78] In June the same year, the Leipzig chemist Arthur Heffter began a series of self-experiments with peyote that concluded in November with the isolation of the alkaloid that produced the visions. Heffter christened it mescaline.[79] In 1898 Yeats and Gonne, in London and Ireland respectively, used 'mescalin' and hashish to attempt a telepathic spiritual union.[80] Neither recorded the result.

*   *   *

By the century's end, the extasias and fantasias of hashish had become a staple of literary and popular culture. In December 1905 *The Strand Magazine* carried an essay titled 'Hashish Hallucinations', a lengthy digest of the experiences described by Théophile Gautier, Bayard Taylor and others, accompanied by the other-worldly illustrations of Sidney Sime. By this time opium, morphine and cocaine, under the newly minted portmanteau label of 'drugs', were firmly established as a social problem, but cannabis was not yet subsumed into this category. The essay relayed its hashish anecdotes without moral censure, with Sime's images locating them in a hallucinatory, proto-surrealist dreamscape. By now *The Strand*'s biggest star, Sherlock Holmes, had a hashish-inspired rival: Prince Zaleski, a detective created by the popular author of fantastic novels and serials M.P. Shiel. Where Holmes entertained himself with his hypodermic needle and 7 per cent solution in his Baker Street rooms, Zaleski's apartment resembled an ancient

24. *Sidney Sime's illustration of Théophile Gautier's hashish*
hallucination, in which Gautier visualised the drug as a jewel sitting
in his stomach.

mausoleum full of treasures, its domed ceiling dimly lit by a green censer lamp:

> The air was heavy with the scented odour of this light, and the fumes of the narcotic *cannabis sativa* – the base of the *bhang* of the Mohammedans – in which I knew it to be the habit of my friend to assuage himself. The hangings were of wine-coloured velvet, heavy, gold-fringed and embroidered at Nurshedabad . . . The general effect was a *bizarrerie* of half-weird sheen and gloom. Flemish sepulchral brasses companied strangely with runic tablets, miniature paintings, a winged bull, Tamil scriptures on lacquered leaves of the talipot, medieval reliquaries richly gemmed, Brahmin gods.[81]

Holmes, the cocainist, was always in vigorous pursuit of his cases, leaping into cabs and trains and materialising unexpectedly in distant corners of Britain. The hashish-smoking Zaleski, by contrast, appeared never to leave his room. He simply issued forth 'a delightful stream of that somnolent and half-mystic talk' into which he was wont to lapse after consuming his fastidiously prepared drug, plucked a few antiquarian volumes from his shelves and solved the most baffling criminal cases using nothing more than his library and the contents of his exquisite mind.[82]

Hashish became a familiar trope in the decadent novels inspired by J.K. Huysmans's *À Rebours*, which the poet Stéphane Mallarmé had characterised as 'an absolute vision of the paradise of pure sensation, which is revealed to an individual when placed before pleasure'.[83] Huysmans's protagonist, Des Esseintes, had been too overstimulated by his aesthetic *mise-en-scène* to experiment with any psychoactive substance stronger than perfume, but several of his successors used it as a major theme. *Hashish* by Oscar Schmitz, first published in German in 1902, was written in the decadent Paris of the *fin-de-siècle* where, as a young poet, Schmitz had been a regular at the symbolist author Rachilde's celebrated Tuesday salon. He had met Jean Lorrain and

Alfred Jarry, immersed himself in the occult works of Eliphas Lévi and attended the rituals of a Martinist group, the same company in which Yeats had been introduced to hashish. He considered the drug to be the perfect stimulus for what he called the '*débauche mentale*', an inner decadence in which morbid fantasies could be indulged without limits but without danger.[84]

Schmitz's novel opens in a Latin Quarter inhabited by dandies, ruined aristocrats and inscrutable wanderers. His protagonist was careful never to take hashish at home: like Jean Lorrain, he feared that uncanny experiences would end up haunting his everyday life. 'You never know what parts of the fantasies will linger in the furniture,' he explains; 'my room must remain pure.'[85] The action shifts to a private hashish club among the bohemian cafés of the Batignolles district, where:

A mild smell of burning resin mingled with the lighter smoke of English cigarettes. On the dark red walls hung deep-black etchings and engravings, in which barely discernible images, like the visions of an incubus, stared down on us. In the corners, between exotic plants, I could make out antique musical instruments like strange reptiles . . . we lowered ourselves down onto pillows. From a table standing between us the count took some tablets of hashish and smiling, offered me the bowl.[86]

The hashish den was a staple of such fictions. It was in part a transposed version of a more familiar literary locus, the opium den, but it also had models in the real world, where the social smoking of hashish was becoming more common in the multicultural cities of Europe and the US. In the American journalist T.W. Coakley's novel *Keef* (1897), the protagonist, a native of Tangier and habitué of its *kif* den, known as the House of Visions, moves to New York. Drifting aimlessly, he wanders into a tobacconist off Union Square to buy cigarettes and notices, from behind a red fustian curtain at the back of the shop, 'the subtle and unmistakable incense of keef'.[87] Smoking parlours, offering *shisha* water-pipes, were often to be found in the back rooms of the city's Turkish tobacconists, and some now offered long-stemmed pipes and *kif* to smoke. There were also, by the 1880s, a small number of private smoking rooms, referred to as 'hasheesh houses' in the true-crime magazines, where they were portrayed as luxurious vice dens for the jaded thrill seeker.

The most detailed account of these dens, 'A Hashish House in New York', by the New York physician and addiction specialist Harry Hubbell Kane, appeared in *Harper's Monthly* in 1883 in the guise of a sensational vice exposé, though it was plainly modelled on Théophile Gautier's fantastical account of the Club des Hashischins. Kane described an excursion to the lower reaches of 42nd Street and a walk through the gloomy warrens of Hell's Kitchen, before knocking on the door of a darkened house that opened to reveal a scene of 'Oriental magnificence'. The spectacle, Kane wrote, 'brought to my mind the scenes of the Arabian Nights, forgotten from boyhood until now'. He and his guide were invited in, asked to exchange street clothes for Oriental finery – gown, tasselled smoking cap and slippers – and offered a *kif* pipe and black lozenges of *majoun* with a gnomic benediction, in this case John Dryden's line 'Take the good the gods provide thee'. When Kane marvelled to his guide, 'Why is everything so magnificent here? Is it a whim of the proprietor, or an attempt to reproduce some such place in the East?' – he received an answer that might have been given by Jacques-Joseph Moreau:

The color and peculiar phases of a hashish dream are materially affected by one's surroundings just prior to sleep. The impressions that we have been receiving ever since we entered, the sights, odors, sounds and colors, are the strands which the deft fingers of imagination will weave into the hemp reveries and dreams, which seem as real as those of everyday life, and always more grand.

This extravagant scene is the prelude to Kane's hashish visions, a lurid fantasia that makes the familiar transit from dream to nightmare in the high style of Fitz Hugh Ludlow. At its conclusion, like Gautier at the stroke of eleven in the Hôtel Pimodan, the author is thrust out onto the grimy streets, dazed and marvelling at 'the cradle of dreams rocking placidly in the very heart of a great city, translated from Bagdad to Gotham'.[88] In his more sober medical writings, Kane reported that the use of hashish in New York was rare, and that its users were a mixture of native New Yorkers and European immigrants, often from Greece and Turkey (the proprietor of his fictionalised hashish house was Greek). Women participated alongside men, sometimes in private side chambers for discretion. Like the larger demi-monde in Paris, New York's drug scene drew a polyglot and cosmopolitan crowd, in which the intoxicants of different cultures mingled: Kane claimed to have been served coca-leaf tea in his hashish den, in Andean style with globe-shaped bowls and silver straws.[89]

In this age of mass global travel and migration, drugs from different cultures had begun to combine in distinctively modern ways. The thrill of discovery is captured in the memoirs of the British engineer James Lee who, after his initiation into morphine and cocaine by his Indian doctor, proceeded through the 1890s and beyond to self-experiment by combining the products of Western pharmacy with the traditional medicines of Asia. The most interesting drug effects, he wrote, came from such combinations, tested judiciously before 'increasing the doses in a carefully thought out system'.[90] He encountered hashish soon enough in the hills of northern Bengal, and found that it could be added to his regimen to create fascinating synergies and potent visionary effects:

> Under the influence of very large doses, combined with a certain proportion of cocaine, it is possible to experience anything one wishes, quite realistically . . . The most pleasing effect, I find, is produced by the mixture of cocaine and hashish, the cocaine injected followed by a dose of the other smoked. The hashish seems to improve the quality of the cocaine, and give it a strange intoxicating effect.[91]

On this combination he often experienced vivid waking visions; if they took on a menacing quality, he discovered that injecting a grain of morphine in solution caused the fear to vanish, 'this being the effect of just the right quantity of morphia following on top of the other drugs'. Under a full moon, staying in a remote hut lent to him by a Naga tribal family with whom he traded rubber, he went for a walk in the jungle and found himself face to face with a tiger:

> We stood staring into each other's eyes for I don't know how long, as time had no meaning for me then. I felt no fear, only curiosity, and I seemed to be able to read its inmost thoughts . . . I took some steps forward, and then I noticed that it was still the same distance from me as before, and I knew that it was a vision. I relaxed and allowed my mind to become passive, and then I noticed that it seemed to be drawing nearer . . . with a concentrative effort, I banished the vision, and returned to my hut, where some pipes of opium brought me back to a normal condition.[92]

In these adventures, Lee considers himself a pioneer. He seems unaware of the echo-chamber of *Arabian Nights* pastiche in which literary accounts of hashish were typically framed, and into which other global travellers like Silas Burroughs reflexively lapsed. In its absence, and with the addition of intravenous cocaine, he employed hashish much as the occultists did: to extend the capacities of the imagination, and allow himself to inhabit and control a visionary inner world. He was resolutely secular in his outlook, and regarded these visions as products of his mind rather than any external or supernatural power, but they also took him on cosmic journeys in which he saw the future evolution of humanity spread out before him in a vast panorama.

Another way to escape the imaginal hold of the Assassins and *The Arabian Nights* was through a more profound immersion in their source. A few years after Silas Burroughs's experiment with *kif* in

25. Isabelle Eberhardt around 1900, during her travels in Morocco and Algeria.

Tangier in 1892, Isabelle Eberhardt, daughter of the Russian nihilist Alexander Trophimowsky, came to live in the city after travelling with her mother from Algeria. They learned Arabic and converted to Islam, and she came of age smoking *kif* nightly in Tangier's cafés.

Eberhardt was determined to become a writer, and to document these secret worlds. She was initiated into a Sufi brotherhood and adopted male dress, allowing her to travel among the nomadic tribes of the largely unexplored interior. She wrote sketches of desert life that were found with her dead body in 1904, after the mud house where she was staying in the remote garrison town of Ain Sefra was washed away during a flash flood. They were finally published in 1920. One of the last pieces she wrote described the time she spent with the *kif* smokers of Kénadsa, in the deep desert at the Algeria–Morocco border, who met each evening in a partially ruined house in the old medina. In contrast with the bejewelled palaces of Western imagination, their aesthetic was one of sparse simplicity: a reed mat, an old chest for a table, a kettle and teapot, a basket of *kif* leaves: 'The little group of *kif*-smokers requires no other decoration, no other mise-en-scène. They are people who like their pleasure.'[93] They gather as the late afternoon sun illuminates the room; one places red-hot coals in the pipes, another plays a *guinbri*, its two strings fastened to a tortoise shell:

> The seekers of oblivion sing and clap their hands lazily; their dream-voices ring out late into the night, in the dim light of the mica-patterned lantern. Then little by little the voices fall, grow muffled, the words are slower. Finally the smokers are quiet, and merely stare at the flowers in ecstasy. They are epicureans, voluptuaries; perhaps they are sages. Even in the darkest purlieu of Morocco's underworld such men can reach the magic horizon where they are free to build their dream-palaces of delight.[94]

PART IV

LOST AND FOUND

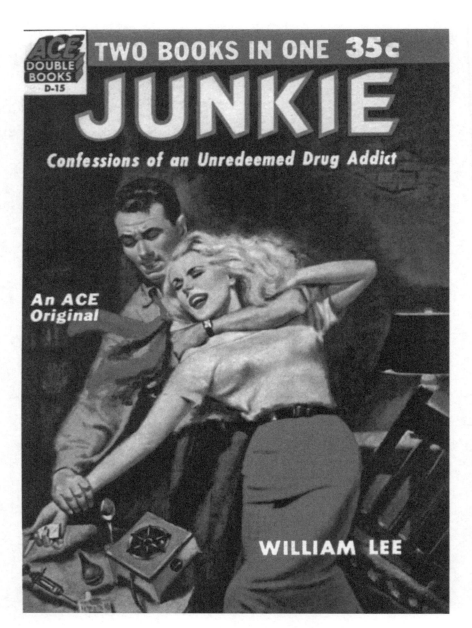

26. *William Burroughs's first novel, written under a pseudonym, presaged the twentieth-century revival of interest in drugs and altered consciousness.*

CHAPTER SEVEN

A SIN, A CRIME, A VICE OR A DISEASE?

'Not an Inebriate', ran a headline in the 23 December 1899 issue of the weekly *Chemist and Druggist* journal. The story was about a young man who had been arrested for breaking the rules of a 'retreat for habitual drunkards' in the London suburb of Twickenham. The patient had been confined to the institution 'to be cured of the habit of taking drugs', but he had absented himself without leave and acquired some cocaine for injection. His defence lawyer argued that that the Habitual Inebriates Act included no mention of 'people addicted to the use of drugs or narcotics'. In the standard printed certificate that bound the defendant, the words 'habitual drinking' had been struck out by hand and 'taking drugs' written in to replace it.[1]

This report from the dying days of the nineteenth century is cited by the *Oxford English Dictionary* as the earliest example of the secondary meaning of the word 'drugs' – today, perhaps, its primary meaning – 'a substance with intoxicating, stimulant or narcotic effects used for cultural, recreational or other non-medical purposes'. Prior to this point, 'drugs' had – as, for example, in the title of the *Chemist and Druggist* – referred generally to all medications. Over the previous decades, intoxicating or mind-altering drugs had often been grouped together, but a qualifying adjective such as 'poisonous', 'inebriating', 'dangerous' or 'addictive' had been required to make the meaning clear. With the twentieth century, the single word 'drugs' came to incorporate all these meanings implicitly. 'Drugs' became a shorthand for a collection of substances that, when not used under appropriate medical supervision, carried the risk of self-poisoning, addiction or mental illness.

The new meaning of 'drugs' arrived largely unnoticed: at first in newspaper headlines, gradually migrating to official documents such as police reports, and by the time of the First World War to common speech. The only official objection to it came from the American Pharmaceutical Association (APA), who were concerned that their legitimate trade was being tarnished. Pressured by their major corporate sponsors such as Johnson & Johnson, the APA complained about the casual use of terms such as 'drug evil', 'drug fiend' and 'drug habit', which they saw as a slur on their entire product range, and urged the press to use more specific terms such as 'narcotics' or 'opiates' when dangerous drugs were involved.[2] But the new usage of 'drugs' caught on, and was soon freighted with meanings and associations that extended far beyond the pharmaceutical trade.

The new century saw the rising tide of what became known as the Progressive Era, a dynamic vision of modernity in which the benefits and risks of drugs were sharply recalibrated. The hallmarks of the new movement were the expansion of government, the growth of citizen and community activism, and the replacement of laissez-faire politics and economics with state intervention and regulation. Despite its name, the great successes of the Progressive Era were achieved by mobilising coalitions from across the political spectrum, uniting progressive and conservative interests against a shared enemy: the overweening might of industry, empire and plutocracy. Alcohol prohibition in the United States, perhaps the era's most ambitious project, was a case in point, achieved by an alliance of progressive forces – temperance campaigners, the women's movement, working men's societies, labour unions and the medical profession – with the conservative voices of the churches and the old Puritan elites. For some, it was a campaign for social justice against the breweries and distillers who were eroding the social fabric for private profit; for others, it was a moral crusade to rid society of the evils of drink. Behind both lay the shared conviction that intoxication, whether by alcohol or other drugs, was running out of control, and threatening the civilised society of the future.

Alcohol was the prime target of this coalition, but the same logic applied to 'drugs'. Both were viewed as insidious manifestations of predatory big business. The trade in addictive medicines was a huge source of revenue for corporate interests, from the alcohol, tobacco and pharmaceutical conglomerates to colonial wealth extractors such as the opium suppliers of British India or the coffee and tobacco magnates of Brazil. Mass production and new manufacturing techniques were making drugs ever cheaper, more pervasive, more potent and more dangerous. Advertising and mass media exploited the vulnerable, selling unattainable fantasies of health while concealing strong drugs in unlabelled patent medicines. The mass of data that now informed public health policy – for instance, the actuarial tables of life insurance companies – showed ever more clearly that life expectancy was shorter and chronic disease more prevalent among heavy consumers of alcohol and other drugs. These were social costs, borne by the public and the economy at large, with all the benefits accruing to big business and private wealth.

*　　*　　*

During the nineteenth century, the dangers of intoxicating drugs rarely went unmentioned. Literary treatments of them almost invariably took a cautionary tone, and medical professionals in particular had become careful to offset any endorsement of their benefits with warnings of their dangers. The prevailing assumption, however, had been that these dangers were best addressed by medical information and social pressure rather than government bans. Outside the borders of the modern West, however, the state prohibition of drugs that began with the criminalisation of hashish in French-administered Egypt had become widespread, most recently in the Philippines, where in 1898 the occupying United States military government prohibited the opium trade. Several

27. The first drug prohibitions were applied to ethnic minorities, such as the 1875 ordinance banning opium in San Francisco's Chinatown.

racially based prohibitions had also been established within the US itself, beginning in San Francisco in 1875 with an ordinance that shuttered Chinatown's opium dens while leaving the sale of opium products in pharmacies untouched.

Among the majority white populations of liberal democracies, meanwhile, individual choice had prevailed. The principle was often formulated in the terms set by John Stuart Mill's *On Liberty* (1859), which addressed the question directly: 'What, then, is the rightful limit to the sovereignty of the individual over himself? Where does the authority of society begin?'[3] Mill's response was that the individual was sovereign over his own conduct; if that conduct led to social harms or law-breaking, such infractions should be punished in their own right, rather than by restricting the freedoms that enabled them. The abuse of freedom, he believed, was properly sanctioned not by law but in the court of public opinion. A person who:

> cannot live within moderate means – who cannot restrain himself from hurtful indulgences – who pursues animal pleasures at the expense of those of feeling and intellect – must expect to be lowered in the opinion of others, and to have a less favourable share of their favourite sentiments; but of this he has no right to complain.[4]

Mill's argument fitted with the widely held opinion that the nineteenth century was the first true era of the individual. Nation after nation had thrown off its imperial yoke, its people choosing to define themselves by their language, community and culture rather than fealty to monarchs or emperors. The rank and file of society had been freed from statutory obligations such as church tithes and religious attendance, and were able to define themselves according to their own lights, as property owners, voters and consumers.

By the early twentieth century, however, the cult of the individual and the loosening of social responsibilities had become the source of gross inequalities and fierce resentment. Progressive thinkers attacked the destructive 'hyper-individualism' that encouraged selfishness and greed. The classic study *Suicide* (1897), in which the founder of sociology Emile Durkheim demonstrated that even the most private acts had a social dimension that could be statistically quantified, diagnosed the twin threats of modernity as 'egotism' and 'anomie': a social condition that left individuals isolated from one another and bereft of the traditional support of community and its established norms. The solution to these problems was the regeneration of civil society. The philosopher John Dewey introduced the term 'social capital' to describe the new communitarian ideal, and US Protestant churches preached a 'Social Gospel' in opposition to the plutocracy and zero-sum economics of the Gilded Age.[5]

From this perspective, those who used drugs were not only voluntarily enslaving themselves to the profiteers of unscrupulous big business, but were mindlessly and selfishly destroying the social fabric. Whether by alcohol or morphine, hashish or cocaine, they were choosing to absent themselves from the shared world of their fellow citizens, abusing their individual freedom by retreating to a solipsistic private reality. Altering consciousness was an extreme form of hyper-individualism; drug users were free-riders who detached themselves from their fellows while still expecting them to pick up the costs of their habit. As these costs were soberly quantified, alcohol and other drugs were increasingly viewed through the lens of their risks, harms and dangers.

The new terms of debate were memorably framed by Norman Kerr, a British physician who in 1884 launched the medical profession's temperance organisation, the Society for the Study and Cure of Inebriety. In the inaugural address, Kerr posed the question of 'whether inebriety is a sin, a crime, a vice or a disease'. If it was a sin, it was a matter for the individual conscience or religious instruction; if it was a vice, it should be dealt with by social sanction, along the lines recommended by John Stuart Mill, and perhaps subjected to government taxation as a luxury. If, however, it was a crime or a disease, it should be addressed with medical or legal controls. No fifth possibility was admitted. Kerr defined inebriety – intoxication by alcohol or other drugs – a priori as a problem; the task at hand was to decide what kind of problem it was. He coined the term 'narcomania' to describe it, and his personal opinion was that 'it is sometimes all four, but oftener a disease than anything else, and even when anything else, generally a disease as well'.[6]

*   *   *

One of the early successes of the Progressive Era was the US Pure Food and Drugs Act of 1906, the federal response to a wave of consumer campaigns against adulterated foods and medicines. This had been dubbed 'The Great American Scandal' by *Collier's Weekly* in a long-

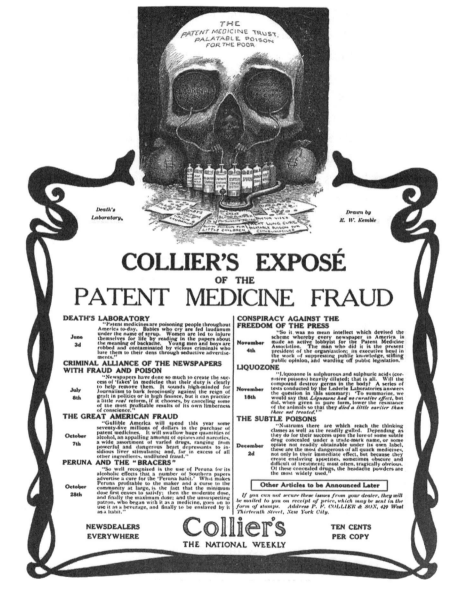

28. *The* Collier's *magazine campaign against unlicenced patent medicines culminated in the US Pure Food and Drug Act (1906).*

running series of exposés, and became a burning issue in 1905 with Upton Sinclair's shocking series of investigations into exploitative and unhygienic conditions in Chicago's meat-packing plants, adapted the following year into his bestselling novel *The Jungle* (1906). The campaign had the professional support of the American Medical Association, at that time controlled by a group of reform-minded doctors who were lobbying for more rigorous licensing of medicine, the expansion of funding for medical schools and a federal department of health. Opposition to the Act was led by corporate lawyers representing the canning, whisky and patent medicine industries. It was steered into law by the unstinting efforts of Dr Harvey Wiley, head of the Federal Department of Agriculture's chemistry division, who would become known as 'the Crusading Chemist'. Wiley was rewarded with the leadership of the new agency established by the Act, the Federal Food and Drug Administration.

The Pure Food and Drugs Act used the term 'drugs' in its broad traditional sense, and was only peripherally concerned with intoxicants, but it strengthened the perception that they should be treated as a public health problem. As well as the use of toxic industrial additives in the food supply, it took aim at the profusion of patent medicines, which by 1900 made up over 70 per cent of the US pharmaceutical market. Chemical testing by Wiley's scientists exposed their unlisted narcotic additives and – equally concerning to temperance campaigners – their high levels of alcohol, in some cases at 50 or 60 proof.[7] The 1906 Act stipulated that all medicines sold in interstate trade must list the drugs they contained, but Wiley saw it as a first step to eliminating the sale of narcotics and stimulants altogether. Even trace levels of cocaine were no longer permitted in patent medicines: as Harvey Wiley told one pharmaceutical attorney, 'the amount makes absolutely no difference'.[8] He envisaged a future in which, as he put it, 'the whole commerce in so-called patent medicines containing these habit-forming drugs would be practically destroyed'.[9] He was firmly opposed to 'drugs' in the new sense of the word, and hoped that his Act would be followed by a federal ban on their trade. Its immediate effect was to remove cocaine almost entirely from public sale and to reduce the trade in narcotic-containing patent medicines by a third.[10]

The Progressive Era's effect on scientific self-experiments with drugs was chilling, and dovetailed with a wider trend in psychology: the shift from introspection to objective and measurable data, and from subjective reportage to group and cohort studies. Emil Kraepelin, the psychiatrist whose taxonomy of mental illnesses underpins the classificatory system that still prevails today, began his career in the Leipzig laboratory of Wilhelm Wundt, measuring the mental effects of *nervina* – 'nerve poisons' – including alcohol, caffeine, chloroform, amyl nitrite and hashish. Despite Wundt's view that intoxicating drugs, like dreams or madness, were a form of inner experience not susceptible to experimental investigation, Kraepelin devised techniques for measuring them. In a series of papers published between 1881 and 1892, he established the changes in reaction time induced by various drugs at different doses, using a newly designed chronoscope that captured reactions to the millisecond.

Kraepelin included himself in these trials, but did not identify himself among the subjects or publish any first-person reportage. He produced a series of tables that correlated drugs, dosage and the number of volunteer subjects with reaction times, reading and number skills, and dynamometer readings of the kind that Freud had used in his cocaine experiments. His results were criticised by some psychologists as inconclusive, with too many uncontrolled variables, and he admitted that the results he elicited 'might be independent of the drug and caused by different influences'.[11] But his methods anticipated the direction in which psychology was travelling, and provided the foundations for a rigorous, statistically based science of psychopharmacology. When he took up a professorship in psychiatry at Heidelberg in 1890, he began to compare drug-induced 'model psychoses', as he named them, with those experienced by mental hospital patients: studies that led to the new disease categories of schizophrenia and manic depression.

These researches convinced Kraepelin of the dangers of alcohol and led him in 1892 to join the ranks of the teetotal. 'To my surprise,' he wrote in his memoirs, 'I found that there was really no reasonable motive for drinking unless one wanted to improve one's mood.' Until this point he had regarded alcohol simply as a fact of social life; now he co-founded a society of abstinent doctors and an alcohol-free public house in Heidelberg, and campaigned to raise awareness of the links between alcoholism and mental illness. It was, in Germany at least, an eccentric position, but one whose time had come:

> The worst problem was that I was constantly involved in endless discussions on alcohol. My strange views were gradually accepted and an increasing number of people assured me that I was quite right and they hardly drank either, only now and again on social occasions. I caused a sensation. I am quite sure that my entire scientific work did not make my name as famous as the plain fact, that I did not drink alcohol.[12]

In 1896 Kraepelin was joined at Heidelberg by a young British researcher, W.H.R. Rivers, who was studying the effect of alcohol and other drugs on mental fatigue. Rivers's work was part of a project to quantify the harms caused by drugs across a broad sample of subjects, in particular their effect on industry and productivity. The 'scientific management' of factory work, established by the mechanical engineer Frederick W. Taylor in 1883, had ushered in the 'time and motion' study, in which industrial processes were broken down into their elementary units and timed with a stopwatch. Rivers hoped that measuring the mental fatigue produced by alcohol and other drugs would not only provide a reliable index of drug effects but allow industrialists to quantify the productivity lost to them.

Rivers was a physician with a keen interest in the 'psychical factors' in medicine: he had written papers on hysteria, neurasthenia and the physiology of vision. He was also a confirmed self-experimenter, who had recently collaborated with his colleague Henry Head on a daring trial that involved severing cutaneous nerves in Head's forearm to measure their regeneration rate. He believed firmly, however, that 'introspection could be made fruitful by the personal experience of a trained observer only'.[13] He adopted Kraepelin's method of 'ergographs', datasets generated by repeated experiments at regular intervals, with variables controlled as far as possible. He noted from his own experience, however, that self-experiment with drugs created feedback loops that were impossible to eliminate. Some trials interested him more than others, and his results were inevitably skewed by 'the interest and excitement produced by taking a substance when the discovery of its effect is the motive of the whole experiment':

> The ergographic curve is an extremely delicate reagent to any form of mental excitement. Any novelty in the course of an experiment may have a very decided effect on the amount of work. The interest of a conversation, the knowledge that the performance is being watched, the view of the weight [on the dynamometer] rising as one works or of the formation of the ergogram on the drum, or any other routine of the daily experiment, may have very obvious effects on the amount of work . . . The interest so aroused will probably be great whether the drug is unknown, so that there is an element of mystery in the occurrence, or whether its nature is known. When in the latter case the subject is himself the experimenter, keenly interested in the possible results of the experiment, this factor of interest must often be very strong.[14]

As Freud had observed twenty years earlier, self-experiment made contradictory demands on scientists: they were obliged to be both subject and observer, a thinking mind attempting to observe its own thought without influencing it. Rivers made dynamometer experiments on cocaine similar to those that Freud had conducted, including himself in the trial and generating similar results. He was unable to demonstrate whether the decrease in fatigue that the drug produced was 'due to its action on the sensory side of the nervous mechanism involved', as Freud had theorised, or simply 'due to the direct action of the cocaine on the muscle', as other researchers had argued.[15]

Unlike Freud, Rivers wrote up his findings exclusively in the third person, relying on passive constructions that disguised his own involvement: his experiments with tobacco, for example, record that 'the smoking was distinctly pleasant, and there was naturally strong sensory stimulation'.[16] He soon moved away from experimental work: on the outbreak of the Great War he joined the Royal Army Medical Corps and was assigned to the work for which he is best remembered, treating the shell shock of Siegfried Sassoon and others at Craiglockhart Hospital in Scotland. But his work on fatigue continued to be widely cited and, along with Kraepelin's, set a benchmark for the quantitative, data-driven psychopharmacology that would dominate the field in the twentieth century.

The retreat from subjectivity was hastened and consolidated by the new doctrine of behaviourism, developed at this time by John B. Watson in his department of psychology at Johns Hopkins University. Watson had a profound personal distaste for the introspective method: as he wrote later, 'I hated to serve as a subject . . . I was always uncomfortable and acted unnaturally.'[17] He found it impossible to align his private responses with those of his colleagues, and gravitated as quickly as possible towards experiments on his well-tended colony of white laboratory rats. In 1913 he published a manifesto for his new approach, 'Psychology as the Behaviourist Views It', in which he redefined the discipline as 'a purely objective natural science':

> Its theoretical goal is the prediction and control of behaviour. Introspection forms no essential part of its methods, nor is the scientific value of its data dependent on the readiness with which they lend themselves to interpretation in terms of consciousness.[18]

Watson rejected the entire notion of consciousness, which he saw as a hangover from the religious past. 'All that Wundt and his students really accomplished', he wrote, 'was to substitute for the word "soul" the word "consciousness"'; but by assuming that 'there is such a thing as consciousness and we can analyse it thoroughly by introspection . . . we find as many analyses as there are individual psychologists'.[19] As behaviourism took hold, self-experiment was driven to the margins. Once the states of consciousness that drugs generated were deemed irrelevant, objective data on their behavioural effects could be sourced equally well from lab rats. The arrival of new instrumentation such as the electroencephalogram (EEG), first used on humans in 1924, further accelerated the trajectory away from studying consciousness and towards measuring brain activity.

Those who persisted with self-experiments became conspicuous outliers. One such was William McDougall, a colleague of W.H.R. Rivers who had travelled with him on the 1898 Cambridge University exhibition to study the psychology of the Torres Strait and New Guinea islanders. McDougall had subsequently made the same transit into experimental psychology, becoming in 1901 a co-founder of the British Psychological Society. In a series of high-profile public debates he defined his position against Watson, whose theory he dubbed 'mechanism', as it treated human beings like machines. McDougall claimed the mantle of the true behaviourist, one who recognised that mental events were relevant to behaviour, which could not be understood without concepts such as desire, striving, valuing or hoping. In 1905 he began experiments to measure fatigue, using a technique later adopted by Rivers in which trial subjects were asked to hit a succession of dots that passed in front of them. When the speed of the moving dots was increased to the subject's maximal rate, fatigue set in quickly and the impact of drugs could be precisely measured.

Like Rivers, McDougall was enlisted during the Great War to treat shell shock victims, after which he teamed up with May Smith, a former student and one of the first female psychologists, for a series of studies funded by the Medical Research Council into the effects of sleep deprivation and drugs on mental functioning. Smith was a gifted researcher, both in laboratory work and in interviewing volunteer subjects, and she earned a reputation for becoming rapidly expert in the skills and trades of the subjects she was studying. Despite being teetotal, she joined McDougall in self-experiments with alcohol, tea, opium and strychnine, alongside control doses of placebos.

This was remarkable enough for the *New York Times* to run an article under the banner headline, 'Scientists Try Drugs on Themselves', reporting that 'More than 100 remarkable drug tests have been undertaken by a woman scientist, Miss May Smith'.[20] 'Scientists Try Drugs on Themselves' was a story that could have been accurately filed on any day in the previous two centuries; it was an irony of the Progressive Era that the belated arrival of a self-experimenting female scientist coincided with the disappearance of the practice, at least from public view.

* * *

In 1900 Jean Lorrain, the *monstre sacré* of decadent Paris, left the city for Nice, where he lived out his final six years as an invalid, a regular patient at the health spas of the Riviera for his tuberculosis, neurasthenia and ether dependency. In 1896 he had been the highest-paid writer in the city, but the arrival of the twentieth century saw his popularity decline; he was harried by lawsuits for libel and defamation, writing furiously but erratically for a public that had moved on. He had often framed his decadence as a hymn to the exhausted final decade of the nineteenth century, a prelude to the terminal and merciful death of civilisation. Just before his death Lorrain met one of his few remaining admirers, the Italian poet Gabriele D'Annunzio, who had traded his life as a decadent poet for one of heroic and frenetic action: turning his ripe, sensual literary style to hymns in praise of war and bloodshed, and trading his Romantic individualism for political alliances with the Futurists and Mussolini. In his new incarnation as a propagandist for Italian nationalism and fearless flying ace, D'Annunzio embraced the new century, the power of industry, the speed of the automobile and the sublime thrill of air travel. Its defining art was the cinema, with its madcap action, jump cuts and perspective shifts; and the drug for D'Annunzio's new incarnation was cocaine, in large quantities, on which he became chronically dependent.

Lorrain left Paris just as the 1900 Exposition Universelle opened, a dazzling showcase of the new century's marvels. Over six months, 50 million visitors had their first glimpses of escalators and moving sidewalks, sound recorders and talking films, planetariums and ocean ship simulators, all presided over by the world's largest Ferris wheel that carried 1,600 passengers at a time. Palaces of Electricity, Optics, Industry, Machines and Agriculture presented spectacular visions of the coming century in the elegant, sweeping styles of Art Nouveau, Belle Epoque, Beaux-Arts and the Vienna Secession. Lorrain's bohemian Paris was being overwritten with the future glimpsed with disdain by Baudelaire: a mass culture of spectacle and sensation, the crowd and the consumer.

Yet Baudelaire's equation of the modern with individual self-fashioning was now itself becoming a thing of the past. If the nineteenth had been the century of the individual, the twentieth was to be that of the collective: group identity, solidarity and mass movements. Baudelaire's legacy became heavily shadowed by the drugs with which he had dabbled. In his memoir of his friend, Théophile Gautier had dismissed the idea that Baudelaire had been ruined by the drugs he wrote about – 'his illness was caused by nothing but fatigue, ennui, sorrow, embarrassments'[21] – but a newspaper retrospective in 1902 added to Gautier's words nonetheless: 'The end of the poet was sad: the fatigues, the troubles, the sorrows, the frequent use of those artificial paradises, opium and hashish, quickened his fall.'[22] The 1900s were hailed as an era of renewal and regeneration after the degeneration and decadence of the 'yellow nineties'. As W.B. Yeats ruefully observed: 'In 1900 everybody got down off his stilts; henceforth nobody drank absinthe with his black coffee; nobody went mad; nobody committed suicide; nobody joined the Catholic church; or if they did I have forgotten.'[23]

Drug users, in the new climate, were hyper-individualists, but they also became a scientifically defined cohort. Emil Kraepelin's experiments with alcohol, together with his personal abstinence, led him to develop the category of 'drinker' as a pathological 'type', statistically at greater risk of mental ill-health and early death.[24] The pejorative associations of 'drugs' obscured the distinctions between different substances, and between occasional and dependent users. Drugs were, in the words of Clifford Allbutt, president of the British Medical Association, ushering in a coming plague of neurotics and neurasthenics 'who scent intoxicants from afar with a retriever-like instinct, and curious in their sensations, play in and out with all kinds of them'.[25] As drug research moved away from subjective descriptions, it became more tightly focused on addiction and its mental pathology. The more drugs were defined in terms of risk, harm and danger, the harder it became to recognise the experiences they generated as a positive contribution to science, philosophy or creativity.

As drug use was pathologised, the popular image of the drug user moved sharply downmarket. Abstinence became a signifier of health-consciousness and social status. In the wake of the 1906 Pure Food and Drug Act, the backlash against industrial food and drug production increased demand for ethical and natural products, and there was a boom in spas and sanatoriums that promised to purge the toxins of everyday life from their patients. In the largest and most renowned, at Battle Creek in Michigan, the medical superintendent John Harvey Kellogg's corn flakes were the signature product of a dietetic system that excluded meat and processed foods along with all drugs including alcohol. Patent medicines were seen as toxic substitutes for the natural therapeutics of fresh air, exercise, electrotherapy and a healthy diet, and any patients who had recourse to narcotic or stimulant drugs were put on strict abstinence regimes.

Since drugs diminished reason and exalted instinct, it followed that their appeal was most dangerous to those whose self-control was weakest: the mentally unfit and the criminal classes. The US physician and addiction specialist T.D. Crothers, one of the first to warn of the medical dangers of cocaine and morphine in the 1880s, now conceived cocainism as a problem of 'the vindictive criminal in lower circles' who 'will stoop to any crime, or any act that promises relief . . . who fights the officer seeking his arrest, who suddenly assaults people on the street or shoots people without any provocation, or sets fire to buildings, commits serious damages, acts wildly and maniacally'.[26] During the nineteenth century, the narcotic or stimulant 'habitué' was commonly stereotyped as a member of the professional or leisured classes: a neurasthenic brain-worker, a dissolute gentleman-about-town or a depressive and socially isolated widow. By the twentieth, these stereotypes had been supplanted by adolescent gang members, petty criminals and, most sharply of all, members of ethnic minorities.

'The problem of the twentieth century', W.E.B. Du Bois observed in 1903, 'is the problem of the color-line', a line that was drawn nowhere more swiftly and firmly than in attitudes to drug-taking.[27] The liberal individualism of John Stuart Mill had always, in practice, been contingent on status, class and race. A drug user of colour, such as Paschal Beverly Randolph, was taking on far greater reputational risks in advocating the use of hashish than, for example, a distinguished medical figure such as Jacques-Joseph Moreau. But the consequences were, in the terms proposed by John Stuart Mill, determined by social mechanisms: Randolph was well aware that his colour made him a target for prejudice, but his use of hashish was not treated as pathological or criminal. In this respect the Progressive Era belied its name, hardening and extending the informal racial discrimination that it inherited.

In the US the most conspicuous example was cocaine. By 1906 it was well advanced on its transit from miracle drug to social menace, and its removal from pharmacy shelves by the Pure Food and Drug Act created an unregulated market for cocaine powder sold on the street in small packets known as 'five-cent sniffs'. This trade was common in the docks, slums and high-crime districts of many US cities, but it quickly became identified with the Black population of the southern states: in 1905 the *New York Times* defined the problem under the headline, 'Negro Cocaine Evil'.[28] Some medical authorities argued that cocaine had a more destructive effect on Black subjects, leading to hallucinations, paranoia and violent impulses, and even that it made them impervious to police bullets. These were the years of Jim Crow, the violent efforts of the white supremacist Southern Redemptionists to roll back the gains of Reconstruction and to suppress the Black vote through literacy tests and poll taxes. Black lynchings reached their grim peak, averaging over a hundred a year in the decade 1891 to 1901. The Progressive Era's ethos of local action and self-empowerment was mobilised by white communities to extend states' rights and allow greater freedom to police forces who cited the dangers of cocaine as grounds for intensifying their control of Black communities. Cocaine dependence was commonly described as 'slavery', a term that positioned anti-drug campaigners and segregationists as the new abolitionists.[29]

*　*　*

The twentieth century's march of mass production, mass culture and mass politics was resisted by a vigorous counterculture of individualism and cultivation of the inner self. In 1904 the sociologist Max Weber began writing *The Protestant Ethic and the Spirit of Capitalism*, in which he argued that the rational pursuit of economic gain left a residue of 'disenchantment': a world haunted by the loss of the sacred, in which the individual was confined in an 'iron cage' constructed by the demands of capital, industry and bureaucracy. As traditional religion retreated, spiritual movements such as New Thought flourished, in which Taoist, Vedic and Buddhist traditions combined with the self-reliant precepts of Emerson and the Transcendentalists and the expansive possibilities of Darwinian evolution to cultivate an inner connection to the divine, the universal mind or the 'infinite intelligence'.

The conviction that human consciousness was expanding or evolving was not limited to Theosophists. In 1901 the Canadian psychiatrist Richard Bucke published his influential *Cosmic Consciousness*, in which he argued that 'higher' or mystical consciousness was becoming more widespread. In 1903 F.H.W. Myers proposed that the subliminal self might include 'ultra-intellectual' or 'supernormal' faculties that a future humanity could learn to harness. In 1909 C.H. Judd, president of the American Psychological Association, used his annual address to oppose the tide of biological materialism by discussing the evolution of consciousness. Judd argued that recent leaps of progress across the arts, science and industry transcended biology, and 'there is no problem of present-day science of more vital importance to the psychologist' than understanding the higher consciousness that had produced them.[30] In 1913 the New York socialite and devotee of New Thought, Mabel Dodge, wrote: 'Nearly every thinking person these days is in revolt against something, because the craving of the individual is for further consciousness, and because consciousness is expanding and bursting through the molds that have held it up to now.'[31]

For this vanguard of the modern mind, however, drugs played no part in the quest for higher consciousness. Like alcohol, they were another padlock on the door of Weber's iron cage. In 1914 Dodge convened an impromptu peyote ritual in her Greenwich Village apartment that went disastrously wrong and threatened scandal; she was horrified that it might be reported as a ' "Dope Party". Horrors! I had heard of such gatherings and they were the antithesis of all I wished to stand for.'[32] Three years later she moved from New York to Taos, where she immersed herself in the spiritual traditions of the Pueblo Indians while campaigning for decades against their use of peyote in their ceremonies. In 1936, when their right to do so was confirmed under the guidelines of the Indian New Deal, she wrote to Harold Ickes, President Roosevelt's Secretary of the Interior: 'Would you stand for hashish, cocaine, or morphine and defend them on the grounds of religious liberty?'[33] 'Drugs', for the spiritually evolved, were chemical agents of control, pumped out by big business to induce a state of mass hypnosis and disconnection from the natural and spiritual world.

Mabel Dodge's guests at her Taos retreat included Aldous Huxley, who provided the defining literary portrayal of this view in his dystopian parable *Brave New World* (1932). Peyote, in the novel, is confined to the 'savages' on the New Mexico reservations; in the civilised society of the future the old intoxicants have been supplanted by the all-purpose palliative *soma*. Alongside other escapist technologies, *soma* has resolved the problems of industrialised mass culture by creating a hyper-mediated fantasy world that wards off the risk of social unrest. Happily distracted, the lower castes of the population toil in factories: 'seven and a half hours of mild, unexhausting labour, and then the *soma* ration and games and unrestricted copulation and the feelies. What more can they ask for?'[34] *Soma* was the sacrament of a political ideology – Huxley had Henry Ford, George Bernard Shaw and the Fabian Society in mind – in which a world state keeps its population in check with a mass culture that allows transcendence only by subsuming individuality. The drug is consumed in ritualised circles, with hymns in praise of Ford and the toast 'I drink to the Greater Being'.[35]

In comparison with his later psychedelic writings, Huxley is curiously vague about the state of consciousness that *soma* produces. It is described variously and at different doses as a euphoric, an antidepressant, a pain reliever and a hallucinogen, but there is virtually no subjective description of the experience it produces. One character, we are told, feels 'a calm ecstasy of achieved consummation . . . she was full, she was made perfect, she was still more than merely herself': satisfaction without achievement, sensation without content, pleasure without a cause.[36] In the drug fictions of the nineteenth century, the narrative often unfolded in the heightened inner world of its mentally altered protagonist; *soma*, by contrast, is 'drugs' through the lens of behaviourism, a chemically induced form of social control that manifests only a blank, delusory euphoria.[37]

Those who advertised their use of drugs in in the Progressive Era did so as an act of ostentatious transgression. The most prominent was the poet, adventurer and occultist Aleister Crowley, who wove them into his public persona in ways the occult societies of the *fin-de-siècle* had fastidiously avoided. The order that Crowley founded in 1907, the A∴A∴, was explicit from the beginning in including drugs as part of its ceremonial and mystical practice. Its handsomely printed journal, *The Equinox*, combined magical tracts and essays with poetry, short fiction and art. Its first issue, published on the spring equinox of 1909, included an article on the pharmacy of hashish by Edward Whineray, the proprietor of Messrs Lowe & Co. pharmacy in London's Bond Street, which advertised in the journal and specialised in supplying obscure drugs, elixirs and incenses to its occult circle. The second issue, on the autumn equinox of the same year, included a flowery Orientalist poem, 'The Opium-Smoker', and a lengthy treatise on 'The Psychology of Hashish' by Oliver Haddo, a transparent *nom de plume* for Crowley, who had been lightly fictionalised under that name in Somerset Maugham's novel *The Magician* the previous year.

Crowley's drug career, he relates here, began in 1898, shortly after he left Cambridge University, when he met Allan Bennett, an analytic chemist, Buddhist and fellow member of the Order of the Golden Dawn. His early experiments with hashish were informed by his reading of Fitz Hugh Ludlow and Charles Baudelaire, and he worked its effects into his magickal practices. Hashish, he wrote, 'loosens the girders of the soul' and produces a marked 'intensification of the introspective faculty', an expansive subjectivity that allows the adept to impose their will and imagination on the subconscious mind and, with further practice, on the external world.[38] Dosage, he discovered, was critical. In small quantities, he described hashish as 'volatile-aromatic':

the thrill, described by Ludlow, as of a new pulse of power pervading one. Psychologically, the result is that one is thrown into an absolutely perfect state of introspection. One perceives one's thoughts and nothing but one's thoughts . . . in other words, in this respect, one possesses the direct consciousness of Berkeleyan Idealism. The Ego and the Will are not involved.[39]

261

At larger doses the effect is quite different, a 'toxic hallucinative' state in which 'the Will and the Ego become alarmed, and may be attacked and overwhelmed'. As adepts from Cahagnet and Randolph onwards had observed, high levels of training are required to combat 'the tide of relentless images'.[40]

Though Crowley was explicit and often boastful about his drug use, he was anxious to steer the reader away from any suspicion of self-indulgence. He stressed the mental discipline required in working with drugs, and described his exacting self-experimental protocols. 'The old Chancery Lane rule', as he described it – referring to the rooms where he was living when he and Bennett began their experiments – was to 'begin with half the minimum dose in the Pharmacopoeia', wait for the effects, and if nothing happens, double the dose. 'If you go on long enough', he advised, 'something is nearly sure to happen!'[41] *The Equinox* was subtitled 'The Review of Scientific Illuminism', and Crowley was straight-faced in presenting his work as rigorous scientific experiment, and himself as a highly trained observer. Just as the microscope revolutionised science, he argued, 'mysticism has revolutionised, again and again, the philosophies of mankind'. In one of the circular Socratic exchanges of which he was fond, he sparred with an imagined scientific inquisitor:

And suppose he retorts, 'You have deliberately trained yourself to hallucination!' What answer have I? None that I know of. Save that microscopy has revolutionised surgery, &c. . . .

Hashish at least gives proof of a new order of consciousness, and (it seems to me) it is this *prima facie* case that mystics have always needed to make out, and have never made out.

But today I claim the hashish-phenomena as mental phenomena of the first importance; and I demand investigation.

[. . .] 'But if you disturb the observing faculty with drugs and a special mental training, your results will be invalid.'

And I reply:

'But if you disturb the observing faculty with lenses and a special mental training, your results will be invalid.'[42]

And so on. It seems unlikely that Crowley persuaded any scientists with these arguments, or indeed expected to, although it may be that the readiness with which disreputable autodidacts could claim scientific justification for their drug use played a small part in the eclipse of self-experiment.

'The Psychology of Hashish' proceeded to discuss concepts from Eastern spiritual traditions, such as *sankhara* and *viññanam*, in the light of hashish intoxication, although here as elsewhere Crowley withheld the practical details of how he incorporated drugs into his ceremonial practices (these are coded and ciphered even in his private notebooks). Over the years he experimented single-mindedly with all the consciousness-altering compounds he could find. He and Bennett made what he described in hindsight as 'fruitless attempts to poison ourselves with every drug in (and out of) the Pharmacopoeia', and by 1910 he was ordering 'anhalonium' – peyote – extracts from Parke, Davis in Detroit.[43] He wrote essays on absinthe, cocaine and ether, all of which he praised in terms that combined euphoria and spiritual attainment:

I have been sucking up the vapour of Ether for a few moments, and all common things are touched with beauty. So, too, with opium and cocaine, calm, peace, happiness, without special object, result from a few minutes of those drugs. What clearer proof that all depends on state of mind . . . Man is a *little* lower than the angels: one step, and all glory is ours.[44]

Crowley developed serious dependencies on cocaine and later heroin, the latter keeping him in its grip until his death in 1947. He loathed his enslavement to them, which made a mockery of his supposedly superior will and forced him into the low-class and criminal

29. Aleister Crowley in 1934, an icon of saturnine hyper-individualism.

company of 'youthful thrill-seekers' whom he despised.[45] His association with drugs was a prominent element in his status as a public
pariah, and a cautionary tale of their necessary end in addiction,
poverty and moral ruin. It was only with the counterculture of the
1960s that his celebration of drugs – along with his unsurpassed talent
for exotic self-portraiture – brought him posthumous recognition as an
icon of saturnine hyper-individualism.

* * *

In 1911 the growing campaign to regulate the global drug trade culminated in an International Opium Convention, held at The Hague in the Netherlands. It included delegations from the United States, Britain and its opium-producing overseas colonies, France, Italy, Persia, Japan, China and Russia, and concluded in January 1912 with a call for all signatory nations to control their domestic trade in opium, morphine, heroin, cocaine and any other dangerous drugs that might emerge in their wake.

In the United States this call was supported by Harvey Wiley's Federal Bureau of Chemistry, and taken up for legalisation by the Democrat representative Francis Burton Harrison, who had served as an adjutant-general with the US Army in the Philippines, where the sale and smoking of opium was banned during the US occupation. The extension of colonial-style bans to American soil was opposed by many, including the American Medical Association, but the corporate pharmaceutical companies who controlled and promoted the narcotics trade were easily equated by advocates of the ban to the brewers and distillers, business magnates accruing vast profits at the expense of public health. The argument was more easily carried with drugs other than alcohol. The majority of the public saw no harm in moderate drinking, but 'drugs' were now firmly identified with problematic types and communities: Chinese or Black ethnic minorities, bohemians, the mentally ill or the criminal classes.

The Harrison Narcotics Act was passed in December 1914, after no more than a few minutes' deliberation, and barely noted among a flurry of other legislation. The following week a resolution for the federal prohibition of alcohol garnered far more attention – during the debate a petition of 6 million signatures, the largest ever presented to Congress, was unfurled from the gallery to the floor – but was narrowly defeated. The Harrison Act added federal authority to the state restrictions on the sale of drugs that had proliferated since the Pure Food and Drugs Act, by making it a federal crime to buy or sell them without registering as a licensed seller and paying business taxes. It also formalised 'narcotics' as a medico-legal term for controlled drugs, expanding its meaning beyond the long-established pharmaceutical definition of sedatives or sleep inducers. Under the Act, cocaine, the most powerful stimulant in the pharmacopoeia, was now classed as a narcotic (Coca-Cola's dena-tured coca-leaf extract, 'Merchandise No. 5', was specifically exempted, the corporation arguing that it was 'used exclusively for flavor').[46] The Board of Health of New York City, where 'hasheesh den' stories were staples of the local newspapers, added hashish to its list of narcotics; the *New York Times* approved, claiming that it had 'practically the same effect as morphine and cocaine'.[47]

By this time informal or illicit drug markets were becoming more visible on the streets of Europe's cities, particularly in ports with large multi-ethnic populations. In Berlin and Hamburg, pharmacies offered morphine injections as a treatment for the nerves; in Paris, where drugs were officially controlled by regulations confining their sale to doctors, packets and vials of cocaine or morphine were sold from table to table in Montmartre cafés. Pharmacies now stocked new 'ethical' products such as Veronal, the first of a new class of sedative drugs, the barbitu-rates. In London, the epicentre of the unofficial drug trade was Soho and the theatreland of the West End, where the Chinese community, the sex trade, dance halls and bohemian jazz clubs all overlapped, creating a subculture where stage dancers and 'party girls' rubbed shoulders with men of colour, and late-night pharmacies sold packets of unlabelled white powders under the counter.

With the onset of the Great War, the newspapers elevated these localised 'drug scenes' into nationwide moral panics. Drug stories fed anxieties about soldiers on duty exposed to unscrupulous peddlers, and equally about temptations to the unprotected women they left behind. In September 1914 rumours filled the British press that soldiers were being corrupted by the cocaine habits of their Canadian and American allies; other exposés detailed how the unlicensed trade was generating profits for German pharmaceutical companies. On the home front, drugs were seen as a threat to young women joining

30. A dealer known as 'Cocaine Emil' selling 5-mark packages on the streets of 1920s Berlin.

the workforce for the first time, and particularly to those mixing in the night-time economies of the big cities.

In France, the drug trade was banned in July 1916 under the Loi du 12 Juillet, which imposed custodial sentences for the possession of opium, cocaine and hashish. The same month, the British government passed similar measures against opiates and cocaine in the Defence of the Realm Act. The Act's primary purpose was to control sensitive information, but it also prohibited the sale of dangerous drugs and imposed licensing hours on the sale of alcohol. The Act put an end to the respectable trade in cocaine and morphine: Harrods, the elite department store, was obliged to withdraw its cocaine and morphine injection kits, advertised as 'A Useful Present for Friends at the Front', and Burroughs Wellcome's bestselling cocaine tablets, 'Forced March', disappeared from the pharmacy shelves.[48] As these outlets closed, the illicit market became more profitable and firmly established, and 'drugs' took on the aura of the criminal underworld in which they now circulated.

On the battlefield, meanwhile, the cocaine sold by neutral Holland, manufactured by the Nederlandsche Cocaïnefabriek from leaves grown in Javanese coca plantations, was supplied to the armed forces of both the Allies and the Central Powers as a performance enhancer, morale booster and remedy against hunger and fatigue, the qualities that had first piqued Sigmund Freud's interest in Theodore van Aschenbrandt's military experiments. It was particularly popular with the dashing new cadre of fighter pilots and flying aces, who used it to stay alert during long flights and sharpen their reactions. As one French military report put it:

Cocaine infused into the few duellists of the air[,] who made use of that cold and thoroughly lucid exaltation which – alone among drugs – it can produce ... at the same time it left intact their control over their actions. It also fortified them, one might say, by abolishing their fear of risk.[49]

It was a role for which alcohol had been used in conflict since time immemorial, but cocaine – and later amphetamines – would prove ideally suited to the challenges of modern warfare.

*　*　*

Sometime around 1910 James Lee returned from Southeast Asia to Britain, this time bringing his Indian wife Mulki with him. As usual, he detoxed from cocaine and morphine during the voyage with his practised method of reduction. He was en route to a new assignment as engineer for a west African gold mine, and decided to take a holiday in London. He found drugs easy to come by, and was soon 'pretty well lit up by frequent doses of cocaine, morphia and cannabis indica'.[50] Normally he stayed indoors when on large doses of drugs, but on this occasion he took an evening stroll down Piccadilly:

The people around me seemed to be abnormal in every conceivable way. Some would appear to be about ten feet high, while others would appear to be microscopic . . . All the time there were spirit-like shapes floating around me, constantly coming and going, and being replaced by fresh ones . . . I strolled along seemingly in fairyland. I was in an intense condition of happiness, without a single care in the world; I seemed to be walking on air.[51]

He wandered into the Underground station at King's Cross and waited in the ticket queue, where he was surprised to see a turbanned Hindu gentleman in front of him. When the clerk asked him what ticket he wanted, he remained silent.

Thinking that perhaps he could not speak English, I spoke to him in Hindustani, asking him where he wished to go, but again no reply.

I next turned to the booking clerk, and said, 'This gentleman, I think, wants a ticket somewhere.'

'Which gentleman?', asked the clerk. 'This gentleman,' I said, pointing to the Hindu, who had not moved.

At once I noticed a strange, scared look appear on the clerk's face, and I put out my hand and felt only air.

'I want a ticket to Piccadilly,' I said hurriedly, putting down the money.[52]

Lee mused on his hallucination as he walked back down Piccadilly under 'a blaze of light from the illuminated signs, and brilliantly-lighted shop fronts', the pavements 'crowded with well-dressed people bent on enjoying themselves'. He was interrupted in his reverie by a young girl whom he immediately recognised as a cocaine user, with dilated pupils, an exhausted expression and 'a pathetic little droop, like that of a tired child'.[53] He accepted the invitation back to her flat on one of the backstreets behind Shaftesbury Avenue, where he asked her how long she had been taking cocaine. She pretended innocence, until he reassured her that he had been using it for years. She expressed astonishment at his good health, and he explained his regime to her. He asked to see her syringe, and inspected the leather washers in the screw caps.

> 'Do you ever take it apart and thoroughly clean it with carbolic or sublimate?', I asked.
> She did not.
> Her system was becoming poisoned by septic matter, while her vitality was being quickly sapped by loss of sleep, and lack of adequate nourishment.[54]

He left the girl with a schedule of instructions to follow, and never saw her again. It was a salutary reminder that not everyone embarked on their drug careers as well prepared as he had been by his doctor in Assam, and that a new generation of drug habitués was growing up beyond the reach of both the law and medical expertise.

Lee shipped out for Lagos, leaving Mulki in England, which she found as exotic as Lee had the Orient. She was quickly bored by the historical sights but thrilled by its modern fairgrounds, amusements and attractions. The highlight of her previous visit had been meeting Queen Victoria, who picked her out of the crowd, spoke to her in Hindi and presented her with a 5-shilling piece, telling her to make it into a brooch. After Lee's assignment in Africa he voyaged to China, returning to join her in London in the autumn of 1914. On 25 November they checked into a hotel in Cheapside in the City of London, and Lee bought some cocaine in solution from Ray's, a pharmacy in Holborn. Back at the hotel they injected it together with some morphine, after which Mulki suddenly lost consciousness. Lee found a nurse, and they gave her artificial respiration and brandy and took her to the nearest hospital, St Bartholomew's. To Lee's horror, the doctor on duty, William Thompson, injected her with another dose of morphine and cocaine, and she promptly died.

In the coroner's report, Thompson attributed her death to 'heart failure — her habit of injection of these drugs cocaine and morphine would quite account for it'.[55] To Lee, with his years of experience, it was clear that Thompson's injection had been a fatal overdose. But for the drug habituée, especially of an ethnic minority, the script was already written. *The Times* reported the story the following day, together with the coroner's verdict of death by misadventure. Lee, it noted, 'had been taking drugs since 1895'; and Mulki, 'like nearly all Indians', was also 'in the habit of taking drugs'.[56] Despite the coroner's verdict, the newspaper report speculated that it might have been an act of suicide.

Lee chose not to relitigate the episode in his memoirs, simply writing with uncharacteristic vagueness, 'Poor girl, she died suddenly in London', from an overdose of 'some drug, I think morphia'.[57] But it marked the end of his drug career:

> When the Dangerous Drug Act came into force, I gave up using all drugs, because the danger and risk of obtaining them was too great. The paltry quantities, about which the authorities make such a fuss, were of no use to me, and I was able to give them up without any trouble or suffering, owing to my experiments and discoveries.[58]

Lee had always been fascinated by the clandestine underworlds in which drug users communed, secluded from the gaze of routine public life. But he had no interest in the criminal milieu to which drugs were now consigned in the modern West. It was not merely the legal status of drugs that had changed in the first decades of the twentieth century, but their meaning. The uninitiated had always associated them with excess, self-indulgence and mental instability, and he had always been discreet in his use of them; but now the rules of the game had changed. He and Mulki had become members of a drug-using class whose habits and lifestyles placed them outside the protection of the law. The adventures in consciousness that Lee had pursued throughout his adult life were now deviant practices, of interest only to psychiatrists and criminologists.

CHAPTER EIGHT

TWICE-BORN

On 25 March 1961, after a gruelling series of forty-three plenary meetings at its New York headquarters over two months, the United Nations (UN) Single Convention on Narcotic Drugs was signed into international law. It was exactly fifty years since the Hague Convention of 1911, the first attempt to impose international controls on the trade, and was hailed by exhausted UN delegates as a long-standing promise fulfilled. The 1961 Convention built on a patchwork of international regulatory agreements – the International Opium Convention signed at Geneva in 1925 under the auspices of the League of Nations, and the establishment in 1946 of a UN Commission on Narcotic Drugs to control the production, import and export of opiate drugs – and it was presented as a natural evolution of these earlier treaties, a tidying-up operation that closed loopholes and brought different jurisdictions into line.

In reality, it expanded the scope and powers of international law considerably. The new convention listed over a hundred different drugs, ranging from synthetic opioids to cannabis, and included new punitive powers that reached far beyond the previous licensing and tax requirements. All signatory nations were (and still are) obliged to suppress the 'cultivation, production, manufacture, extraction, preparation, possession, offering, offering for sale, distribution, purchase, sale, delivery on any terms whatever, brokerage, dispatch, dispatch in transit, transport, importation and exportation of drugs contrary to the provisions of this Convention'; and all offences against the new laws to be met with 'adequate punishment particularly by imprison-ment or other penalties of deprivation of liberty'.[1]

Many of the convention's provisions had already been introduced in the USA by the Federal Bureau of Narcotics, established in 1930 after the collapse of alcohol prohibition. While alcohol had proved too deeply rooted in mainstream society to eliminate, the drug prohibition established by the 1914 Harrison Act became ever more firmly entrenched. The Federal Bureau of Narcotics was the fiefdom of its commissioner Harry Anslinger, who built up its power and influence with eye-catching propaganda campaigns against the 'drug menace' that identified drugs with marginal communities regarded by the general public as threats to the social order. Anslinger extended his remit by pushing for further drug controls, notably the Marijuana Tax Act of 1937, which criminalised the trade in cannabis by unregistered sellers. Registration was confined to a handful of medical bodies and to apply for it was, *de facto*, to incriminate oneself as an illicit supplier.

Under Anslinger's dispensation, drug use became both a crime and a vice. It was policed in tandem with other criminal vices, such as the sex trade and illicit gambling; these were crimes in which perpetrators and victims were willing accomplices, and the 'vice squads' tasked with policing them were often obliged to resort to entrapment and bribery, and became pawns in rivalries between criminal gangs, just as they had during alcohol prohibition. Where drugs had been viewed during the Progressive Era as a tool of profiteering big business, now they became products of organised crime. Anslinger's campaigns identified them with ethnic minorities: he used the foreign term 'marijuana' in his public statements to obscure the fact that the plant had long been familiar as cannabis, the medical term, and hemp, the fibre crop. 'The menace of marijuana is comparatively new to the United States,' he wrote in 1937. 'It came in from Mexico, and swept across the country with incredible speed.'[2]

Anslinger reached further back into history to summon the legends of the Assassins, 'whose history is one of cruelty and murder'. 'Its members are confirmed users of hashish', he wrote, under the influence of which they 'engage in violent and bloody deeds'. One of his campaigning articles, 'Marijuana: Assassin of Youth', claimed that 'no-one knows, when he smokes it, whether he will become a philosopher, a joyous reveller, a mad insensate, or a murderer'.[3] Anslinger began pushing for an international treaty in 1948, and his reports from this era fed into the evidence presented to the 1961 UN Convention along with those of Charles Vaille, the French representative of the Commission on Narcotic Drugs, for whom the Algerian war of independence gave the Assassin myth a renewed potency. In the run-up to France's 1970 Drug Law, Gaullist politicians recalled the association between hashish and 'assassin', and warned of 'foreign elements' infecting the youth with 'artificial paradises'.[4]

Perhaps the most striking linguistic flourish in the 1961 Convention is its description of 'addiction to narcotic drugs' as 'a serious evil for the individual', which is 'fraught with social and economic danger to mankind'. The convention affirms that its signatories are 'conscious of their duty to prevent and combat this evil'. Drug use, in this formulation, is not merely a crime, vice or disease but a sin. The term 'evil' is most unusual in UN documents, which are careful to avoid inflammatory language: it is not used to describe slavery, torture or nuclear war. Even the genocides of the Second World War are described in the Universal Declaration of Human Rights simply as 'barbarous acts', and in the Genocide Convention as an 'odious scourge'.[5] Its use in the Single Convention on Drugs echoes that of terms such as 'opium evil' in the anti-drug campaigns of the nineteenth century, and appears to have been incorporated from the evidence submitted by the Roman Catholic Church.[6] The use of the word 'evil' framed drugs as an existential threat to the international order, and placed the new laws beyond the reach of evidence or political debate.

*   *   *

The 1961 Convention remains the foundation of today's international drug control regime which, ten years later, President Richard Nixon would brand with the enduring slogan 'The War on Drugs'. Yet 1961 was also the moment when the meaning of drugs underwent a profound and irreversible shift from the Progressive Era views that the convention enshrined. 'The sixties', as a cultural phenomenon, was still in embryo; social historians commonly date its commencement from the death of John F. Kennedy in 1963 – the year, according to Philip Larkin, that sexual intercourse began, 'between the end of the "Chatterley" ban/and the Beatles' first LP'.[7] This was a watershed moment for drugs, just as it was for sex and rock'n'roll. The early years of the decade were Janus-facing ones, in which the version of modernity that coalesced around 1900 was asserting itself against a new set of values in which drugs were no longer an evil, but the portal to novel experience and expansive horizons of human possibility.

In 1961 London's Metropolitan Police noticed for the first time that cannabis, until this point a 'vice' problem confined to the city's ethnic Caribbean and Indian population, was spreading into the social mainstream:

> Of the white users of this drug they are mainly in their late teens or early twenties and are of the type frequenting the jazz clubs and coffee bars of the West End where this activity is mainly confined. There are signs, however, of it spreading to whites of similar tendencies in Brixton, Kensington, Chelsea, Paddington and the Notting Hill area . . . the most dangerous trend is the interest shown by irresponsible young white people of both sexes in this drug.[8]

The most alarming example was the arrest for cannabis possession of Godfrey Peter Manley Glubb, the twenty-year-old son of Lieutenant-General John Bagot Glubb, known as 'Glubb Pasha', a highly decorated bastion of the military establishment who had commanded the Arab Legion in Transjordan during the 1948 Arab–Israeli War.

This was a striking example of what sociologists had recently dubbed 'the generation gap', but it was also emblematic of a moment when the balance of geopolitics was shifting dramatically. In November 1961 the United Nations also established its Special Committee on Decolonisation, and Britain was in the process of surrendering Nigeria, Tanzania, Uganda, Kenya, Malawi and Zambia to their own populations. The post-war generation who birthed the drug culture of the 1960s had been raised, in contrast to their parents, as consumers in a rapidly globalising economy and were discovering that alcohol was not the world's only intoxicant. The drugs that their parents had considered the 'degenerate habits' of 'inferior races' held no such stigma for them; on the contrary, they held the promise of exotic adventure and self-discovery.

In 1961 in Morocco, the American author Paul Bowles was at work on the stories that reintroduced the *kif* smokers of the Maghreb to Western literature. Bowles settled in Tangier in 1947, and his best-selling 1949 novel *The Sheltering Sky* brought attention to both him and his Saharan locales, with their combination of flyblown squalor and searing natural beauty. After Morocco achieved independence from France in 1956, he became a magnet for literary visitors – Tennessee Williams, Gore Vidal, Truman Capote, Allen Ginsberg and William Burroughs – and travelled widely in the desert interior. He received funding from the US Library of Congress to record the country's traditional musicians and its street poets and storytellers, and immersed himself in its underworld of *kif* smokers, to the disapproval of Morocco's Ministry of the Interior.

The picture of *kif* that emerged from Bowles's stories, published by the doyen of beatnik publishers, San Francisco's City Lights, in 1962 as *One Hundred Camels in the Courtyard*, bore little resemblance to the terrifying 'marijuana' of Harry Anslinger, nor indeed to the *Arabian Nights*-inspired scenes portrayed by the West's nineteenth-century travellers and romancers. Bowles wrote in an evocative but simple prose, largely shorn of introspection; *kif* was smoked on almost every page, but its mental effects were rarely described directly. It was a presence, as he explained, which linked scenes and memories that were not obviously connected except by '*kif*-directed motivations' that 'forced [them] into a symbiotic relationship'. As Jacques-Joseph Moreau had observed in Egypt a century earlier, its widespread use promoted a shared culture in which coincidences, dreams and supernatural events were woven into waking life. Moroccan *kif* smokers, Bowles wrote, inhabit two worlds: one of everyday reality, and another 'in which each person perceives "reality" according to the projections of his own essence'.[9]

Kif inhabited a similarly ambivalent role in Moroccan daily life. It was illegal, but the law was only sporadically enforced and it was easily available from market sellers. One might be searched in the street and arrested for possessing it, but it could be safely smoked in certain cafés owned by members of the royal family. Bowles's stories immersed the reader in the minutiae of life among the poor and marginalised, and followed the narrative rhythm of Moroccan folk tales, but he was also using *kif* as a tool for modernist literary experiment, to juxtapose unrelated scenes and parallel streams of consciousness. Often these experiments would reveal surprising truths, just as, he suggested, the Moroccan *kif* smoker's apparently aimless daydreams could equally be seen as 'a pilgrimage undertaken for the express purpose of oracular consultation'.[10]

Bowles translated the travel writings of Isabelle Eberhardt, left unpublished on her death in 1904. He also brought the voice of the Moroccan *kif* smoker, absent from the hashish literature of the nineteenth century, to his new generation of Western readers. Among the local storytellers he listened to, recorded and translated was Mohammed Mrabet, a Tangier local and regular visitor to Bowles's apartment where he cooked, smoked *kif* and told tales of fellow outsiders, drifters and *kif* smokers that emerged both from real-life events and from his dreams. Bowles recorded and published Mrabet's tales, and their cadences and arabesque turns informed his own writing. The emerging drug culture,

31. Paul Bowles with Mohammed Mrabet, whose hashish tales Bowles translated into English.

like its nineteenth-century predecessor, would be popularised and chronicled by predominantly white, Western and male voices, but from its beginnings it unfolded in a postcolonial modernity, and was alive to a wider spectrum of lived experience.

In New York in 1961, a short cab ride from where the UN delegates assembled to discuss the new global drug controls, the Dollar Sign coffee house on the Lower East Side was selling peyote, bought by mail order from a wholesale supplier in Laredo, Texas, powdered and sold in gelatine caps for 50 cents each. At the San Remo café in Greenwich Village, haunt of Allen Ginsberg, Miles Davis and Jack Kerouac, the writer Terry Southern recalled that 'people started chopping them up and eating them like figs'.[11] During the Progressive Era, bohemian enthusiasts for Indian culture such as Mabel Dodge had led campaigns against the indigenous use of peyote, regarding it as a scourge of the tribes comparable to bootleg whisky. In the coming version of modernity, it would attract a new generation to Native American culture, spirituality and political rights.

The rediscovery of non-Western drugs was one symptom among many of a generational shift away from the social movements of the Progressive Era, and towards something more closely resembling the age of the individual that had preceded it. The communal and social movements of the 1900s had devolved, for many, into a soul-crushing culture of mass conformity. The screenwriter and critic Ben Hecht, in his memoir *Child of the Century*, published in 1954, characterised his lifetime as an era systematically dedicated to the 'defeat of human individuality':

> Our civilization is like an implacable advertising campaign determined on overcoming all individual expression . . . The individual healthy enough to think his own thoughts is numbed by the pandemonium around him. He must seem not an individual but the most rabid of anarchists if he would 'think for himself' . . . Individual man is being remade as a faceless part of Democracy, Socialism, Communism and Fascism.[12]

These criticisms were by now well established in the psychological sciences, where a reaction against the tenets of behaviourism was under way. The counter-tradition of psychoanalysis, which placed basic human drives and individual life experience at the centre of its inquiries, was buttressed by sociologists and anthropologists who believed that Nazi Germany had demonstrated the necessity for a democratic state to encourage citizens to think for themselves. In 1950 the doorstopping, multi-authored survey *The Authoritarian Personality* from the University of California at Berkeley, funded by the American Jewish Committee, introduced an influential scale for assessing an individual's susceptibility to fascism, setting subservience to authority against the creativity, individualism and open-mindedness that were required to nourish a free society. By the early 1960s these values were being embraced by a post-war youth generation who, living in the shadow of the nuclear bomb, found common cause in their distrust of the state and the conformist suburban culture it valorised. As the social historian Theodore Roszak recalled in his classic account of the process, *The Making of a Counterculture* (1968), a continuous succession of flashpoints and 'issues' marked out the new sensibility from the old:

> By the late fifties, bolder members of this generation in identity crisis had already decided that Beatnik poets and Greenwich Village folk singers were better models than fathers who have sold their soul to General Motors or mothers who racked their brains all day to bake a better biscuit. They dreamed of being 'on the road' rather than on the job. Hair became an issue: growing it long meant dissent. Dirty words became an issue: the expletive undeleted meant irreverence. Sex out of marriage became an issue: it meant risk-free promiscuity and possibly women out of control. Puffing a joint became an issue: it meant courting outlaw status and outlaw consciousness. It was easy to find issues. They grew plentifully in the cracks between the dying morality of privation and a lush new economic order that all but made gratification mandatory.[13]

The catch-all term 'drugs', still freighted with Anslinger-era stigma and applied with equal moral and legal force to cannabis, peyote and heroin, was ripe for reclamation. In Roszak's analysis, the two trump cards held by the emerging counterculture were the poetry of the beats and LSD. A conspicuous trailblazer was the writer William Burroughs, who by the mid-1950s was busily assembling a constellation of mind-altering substances from sources that ranged from discarded pill bottles in bathroom cabinets to hallucinatory jungle vines in the Colombian Amazon. Burroughs's first novel, *Junky* (1953), cast a jaundiced eye on the emerging beat scene as he hunted down his desired substances among the hustlers and petty criminals of New York, New Orleans and Mexico City. His unsparing descriptions of his drug experiences, balanced between dispassionate observation and waking dream, recalled the subjective reportage of the nineteenth century:

> Morphine hits the backs of the legs first, then the back of the neck, a spreading wave of relaxation slackening the muscles away from the bones so that you seem to float without outlines, like lying in warm salt water. As this relaxing wave spread through my tissues, I experienced a strong feeling of fear. I had the feeling that some horrible image was just beyond the field of vision, moving, as I turned my head, so that I never quite saw it. I felt nauseous; I lay down and closed my eyes. A series of pictures passed, like watching a movie: A huge, neon-lit cocktail bar that got larger and larger until streets, traffic, and street repairs were included in it; a waitress carrying a skull on a tray; stars in the clear sky.[14]

Burroughs was drawn to memoirs of the criminal and drug underworlds of the past almost as keenly as to drugs themselves: *You Can't Win* (1926), the burglar and hobo Jack Black's memoir of safe-cracking, vice, prison and life on the road, was an enduring favourite, as was James Lee's memoir, *The Underworld of the East*. Burroughs praised Lee's book in an unpublished foreword as a narrative 'conjured from the unpolluted air of the nineteenth century' before drugs became taboo, when their pleasures could be celebrated frankly. He was fascinated by the details of Lee's drug reduction and withdrawal system, and his 'wholesome regime of four daily injections of morphine and cocaine in equal quantities, four pipes of opium before retiring, and hashish throughout the day'.

Lee's 'happy and nostalgic book' became a charter and a template for Burroughs's rediscovery of a polydrug regime from similarly eclectic sources: medical expertise, street pharmacology and foreign travels.[15] At the end of *Junky*, in Mexico City, he noted that 'peyote is a new kick in the States. It isn't under the Harrison Act, and you can buy it from herb dealers through the mail.'[16] He sourced some from a local supplier, examined it closely, peeled four heads and grated the pulp 'until it looks like avocado salad'. He swallowed it down with difficulty, vomited and waited for the effects. 'Peyote high is something like a Benzedrine high,' he concluded. 'You can't sleep and your pupils are dilated. Everything looks like a peyote plant.'

Benzedrine was an early addition to the street pharmacy, encountered through hitching all-night lifts with long-distance truck drivers. In 1954, on the run from legal problems and the manslaughter of his wife, Burroughs was inspired by Paul Bowles's writings to visit Tangier, where he included hashish in his daily repertoire. He became captivated by the Assassin legend, which he twisted into a countercultural myth of initiation into an underground drug culture, much as Théophile Gautier had in his accounts of the Club des Hashischins. The Old Man of the Mountain, Hassan-i-Sabbah, became for Burroughs a recurring motif of resistance to authoritarian control. Where Jacques-Joseph Moreau had dosed his subjects with the faux-Assassin blessing 'This will be deducted from your share in Paradise', Burroughs identified Hassan-i-Sabbah's creed with the motto 'Nothing is True. Everything is Permitted.'[17]

* * *

In February 1961, while the UN's deliberations over the Single Convention on Drugs were in progress, the most consequential self-experimental programme of the era was launched on the suburban fringes of Boston. Beside the fireplace in his rented home in Newton, looking across the river to Harvard University, the research psychologist Timothy Leary and his senior colleague Richard Alpert, assistant professor in clinical psychology, each counted out five 2mg pink pills of Indocybin, the synthetic compound psilocybin recently isolated from a Mexican mushroom and brought to market by Sandoz Pharmaceuticals of Basel, Switzerland.

The previous summer, Leary had made his first experiment with psilocybin mushrooms in a villa overlooking a golf course in Cuernavaca, a moneyed resort an hour south of Mexico City. His luxury poolside holiday devolved into a mind-shattering experience during which he laughed and cried helplessly while being whisked through time and space on a tour of palaces, temples, jungles and kaleidoscopic vortices. After five hours he returned, exhausted and exhilarated, with a conviction that set the course for the rest of his life. On his return to Harvard, he approached Alpert to join him in forming the Harvard Psilocybin Project, which aimed to extend his self-experiment to a cohort of faculty members and graduate students. Between them they proposed to assemble a dossier of subjective reports that would, they hoped, open up a new and exciting field of psychology.

Despite the eclipse of introspection during the Progressive Era and the heyday of behaviourism, self-experimentation had not entirely disappeared. The method retained a handful of stalwart champions across the sciences, though they found themselves pushing against the grain. One prominent example was the biologist J.B.S. Haldane, whose father John Scott Haldane was a heroic scientist in the nineteenth-century mould: in 1895 he had himself lowered into a well in which five men had died from inhaling unidentified poison gas, and identified the presence of hydrogen sulphide. J.B.S., in the course of a high-profile career that ranged across genetics, mathematics and physiology, made experimenting on himself a point of principle, and cheerfully admitted to having on many occasions 'lost consciousness from blows on the head, from fever, anaesthetics, want of oxygen and other causes'.[18] He regarded animal subjects as 'useful for rough experiments' but distinctly second-best when the aim of the research was to study human responses.

For Haldane, self-experiment was not merely the clear ethical choice – in their own researches, scientists should be prepared to lead from the front – but provided superior results. In 1927, in a celebrated essay entitled 'On Being One's Own Rabbit', he described a series of experiments in which he 'wanted to find out what happened to a man when one made him more acid or more alkaline'.[19] He attempted to saturate his blood with carbon dioxide by overbreathing, with a colleague on hand to prod him when he passed out; by eating 3 ounces of bicarbonate of soda; by drinking a pint of corrosive hydrochloric acid solution; and by eating ammonium chloride. He pronounced himself 'quite satisfied to have reproduced in myself the type of shortness of breath which occurs in the terminal stage of kidney disease and diabetes'.[20] 'It might be thought that experiments such as I have described were dangerous', he concluded, but 'This is not the case if they are done with intelligence . . . Experiments in which one stakes one's life on the correctness of one's biochemistry are far safer than those of an aeroplane designer who is prepared to fall a thousand feet if his aerodynamics are incorrect. They are also perhaps more likely to be of benefit to humanity in general.'[21]

Haldane's practices were rational, courageous and hair-raising in equal measure, and in part an assertion of the heroism of his father's generation against a bloodless twentieth-century scientific method. He regularly volunteered for human trials by other scientists, including in 1949 with the pharmacology department of University College London, where he attempted to complete cognitive tasks under the influence of nitrous oxide in a project supervised by the young psychopharmacologist Hannah Steinberg. Steinberg went on to have a distinguished career studying the actions of drugs which, following Haldane, she made a point of testing on herself. Self-experiment, she wrote, informed her insights into how much their effects could vary, depending on the subject and their emotional state.[22] Her method made her an outlier among her colleagues, but her gender attracted rather less attention than May Smith's had a generation previously. In 1970 she became the first professor of psychopharmacology at University College London.

It was within pharmacology, as in Steinberg's case, that self-experiment had remained most deeply rooted, particularly when new psychoactive drugs were being synthesised and tested. The discovery of amphetamines, the most significant family of psychoactive drugs to emerge in the early twentieth century, illustrated its benefits. In June 1929, at the physiological department of UCLA, the US chemist Gordon Alles developed a novel compound from adrenaline, beta-phenyl-isopropylamine, which he trialled on guinea pigs. He learned from his lab animals that it raised blood pressure, but self-experiment told him far more than they ever could. When injected with it by a colleague, he recorded 'a feeling of well-being' and became talkative; he invited his colleague home to dinner with his wife, at which his conversation was unusually sparkling. Afterwards he had a 'rather sleepless night' in which his 'mind seemed to race from one subject to another'.[23]

Alles and his colleagues self-experimented with daily doses of the compound and named it amphetamine. It increased their energy, excite-ment and 'pep', and in some cases their anxiety. Alles created further compounds in the lab, and on 16 July 1930 he took 126mg of a variant he had created by adding hydrogen atoms to the molecule. This produced more marked visual effects: the room seemed 'to be filled with curling smoke'. 'Cannot think in long chains,' he scribbled; 'thought skips rapidly but seems quite clear.' These changes were accompanied by a 'generalised feeling of well-being', along with dilated pupils and grinding teeth.[24] The compound would become known as methylene-dioxyamphetamine or MDA, a close cousin of MDMA (ecstasy).

Alles was granted a patent for amphetamine and took it to the phar-maceutical company Smith, Kline and French (SKF), who in 1934 brought out Benzedrine, a volatile preparation of amphetamine in a metal tube for spraying into the sinuses as a decongestant. It was an instant commercial success. SKF were equally interested in its potential as a mental performance enhancer, and embarked on a series of trials with subjects ranging from fellow chemists to children with learning disabilities. Advertisements in medical journals offered free samples of Benzedrine sulphate pills to doctors, who reported that it could be usefully prescribed for fatigue, depression and narcolepsy.[25]

It was crucial, however, to demarcate this type of freewheeling professional experiment from indiscriminate use by the public at large. Among the early recipients of Benzedrine was William Sargant at London's Maudsley Hospital, an early and enthusiastic champion of psychiatric treatments such as insulin coma and electroshock therapy. Sargant noted amphetamine's early promise in the treatment of depressive patients, and proceeded to experiment on himself:

> Though I have rarely taken drugs for experimental purposes, lest they might bias my judgment, I did try one of these tablets one Saturday afternoon, then I walked energetically around the Zoological Gardens with a most delightful sense of confidence and not in the least fatigued. Returning to the hospital, I worked hard all that evening, still happy and vigorous. It suddenly occurred to me that, unless this top-of-the-world feeling were due to some other cause, Benzedrine should clearly help me to pass examinations.[26]

Sure enough, Sargant passed his Diploma of Psychological Medicine on amphetamines with flying colours, and an informal small-scale trial with his Maudsley colleagues established that it 'noticeably improved the percentage of correct answers'. Sargant theorised that this was connected with its virtues as an anti-depressant: it did not necessarily boost intelligence, merely confidence. A few years later, however, he was appalled to discover that Benzedrine was being bought from pharmacies 'in the form of tablets such as "Drinamyl" or "Purple Hearts", which psychopaths, drug addicts and simple delinquents used as a source of cheap "kicks"'. Although 'I had myself tested its effects beforehand and found that it agreed with my particular constitution', he lobbied vigorously for it to be limited to medical prescription.[27] The lines between research and recreation, science and advocacy, professional and personal use were more clearly drawn in the twentieth century than in Sigmund Freud's day, as Leary and Alpert would quickly discover.

*   *   *

On 31 January 1960, the author, poet and classicist Robert Graves joined Gordon Wasson and his wife, Valentina, in their Manhattan apartment overlooking the East River for an experiment with 'magic mushrooms'. Wasson, a senior public relations man with the investment bankers J.P. Morgan, had caused a sensation in 1957 when his article 'Seeking the Magic Mushroom' was splashed across eight vividly illustrated pages of the May 1957 issue of *Life* magazine, in which Wasson described his participation in sacred mushroom rites in the remote mountains of Oaxaca, Mexico. It was Graves who had originally alerted Wasson to the story, on the basis of a paper by the Harvard ethnobotanist Richard Evans Schultes, written in 1939, which identified the sacred plant described in Aztec codices as *teonanácatl*, 'flesh of the gods', as a hallucinogenic fungus.[28] Schultes's work had aroused little interest outside his specialist discipline at the time, but by the mid-1950s it was of huge significance to enthusiasts such as Wasson, who had developed an amateur but tenacious and well-funded fascination with the sacred and mind-altering role of mushrooms in prehistory.

Wasson offered Graves the mushrooms 'in crystalline form', the Indocybin that the Sandoz Pharmaceuticals chemist and discoverer of LSD, Albert Hofmann, had isolated in 1958 from Wasson's Mexican mushroom specimens, and with which Leary and Alpert would make their first formal experiment the following year. Graves was unsure of what to expect and feared he would be visited by 'hideous demons and nameless horrors', but as the pink pills took effect he achieved an effortless transcendence:

> My closest experience to this had been in early childhood when, after waiting endlessly in the cold, dark hall, my sisters and I saw the drawing-room door suddenly flung open and there blazed the Christmas tree: all its candles lighted, its branches glistening with many-coloured tinsel.[29]

Wasson played his tape recording of Maria Sabina, the Mazatec *curandera* who had guided him through his first mushroom encounter, invoking Christ in the form of Tlaloc, the pre-Hispanic deity of rain and fertility. For Graves, 'it might have been the Goddess Aphrodite addressing her forward son Eros'. He was entranced, and 'the song-notes became intricate links of a round golden chain that coiled and looped in serpentine fashion among jade-green bushes'. After several hours the trance and the visions faded, and as the company ate cold turkey sandwiches in the kitchen Graves realised that 'a curious bond of affection had been established between us: so strong that I felt nothing could ever break it'. He parted the next morning, 'profoundly refreshed and (in Wordsworth's phrase) trailing clouds of glory'.[30]

Reflecting on the experience, Graves concluded that it had been more than a personal transformation; it had changed the meaning of 'drugs' entirely:

> The word 'drug', originally applied to all ingredients used in chemistry, pharmacy, dyeing and so on, has acquired a particular connotation in English, which cannot apply to *psilocybin*: 'to drug' is to stupefy, rather than to quicken the senses . . . This particular virtue of *psilocybin*, the power to enhance personal reality, turns 'Know Thyself' into a practical precept.[31]

In the pivotal years of the early 1960s, many things changed at once. At the moment when the Progressive-Era notion of 'drugs' was enshrined in international law, a new class of drugs emerged under the name 'psychedelics', a coinage that aimed to escape and overwrite the negatives of 'drugs' with positive associations that aligned it with the latest iteration of modernity. Graves was among the first wave of prominent scientists, writers and intellectuals who had been repelled by the idea of 'drugs' as they had previously conceived them, but became outspoken and passionate advocates for what became known as 'the psychedelic experience'. This was, in turn, part of a broader shift in which the behavioural and normative psychology that underpinned the Progressive Era's rejection of drugs was challenged by a rediscovery of mystical experience and, in particular, the writings of William James and the capacious view of mental life he had advocated. All these shifts combined to create new possibilities for drug-induced states of consciousness and even, through its reframing by influential figures such as Wasson and Graves, to change the experience itself.

The curious fact that drug effects could be transformed by their social context had recently been examined by the sociologist Howard Becker in his 1953 paper 'Becoming a Marijuana User'. Becker had studied jazz musicians in New York, among whom cannabis-smoking was common; this was, he concluded, not because its members were deviants or criminals, but because they enjoyed its effects in ways that were not available to the straight majority. 'Marihuana use', he wrote, 'is a function of the individual's conception of marihuana and the uses to which it can be put.' Most Americans were not acquainted with cannabis users, and the experience of getting high would be strange, alienating and disturbing for them; but within a subculture that shared and celebrated 'concrete referents of the term "high"', a novice could quickly be initiated into recognising its pleasures. 'The new concepts', Becker argued, 'make it possible for him to locate these symptoms among his own sensations' and to associate them with enjoyable effects. Light-headedness and confusion, rather than being perceived as symptoms of illness, would become cues for giggling and verbal flights of fancy, and anxiety or paranoia could be shared in a spirit of camaraderie and amusement.[32]

Becker's insights were equally applicable to the process by which mind-altering drugs had found their niches in the nineteenth century, and they were timely at a moment when psychiatry's biomedical turn was promoting the assumption that their effects were constructed entirely in the brain. His theory was supremely applicable to psychedelics, as the ideas introduced by Graves and the Wassons demonstrated. In the years after psilocybin had been isolated from the Mexican mushroom and identified as a psychedelic, it gradually became clear that psilocybin-containing fungi were indigenous across Europe and the United States. There was no documented cultural or religious tradition for their use, but there were many examples of accidental ingestion, recorded over the years by doctors, mycologists and toxicologists. None of these had been conceived as a mystical or spiritual experience. Their subjects, on noticing the early onset of psilocybin's effects – dizziness, gastric disturbance, odd and intrusive thoughts – had typically leapt to the conclusion that they had inadvertently eaten poisonous fungi and were undergoing a toxic crisis, perhaps a fatal one.[33] Hallucinatory effects and visual distortions were experienced as delirium or fever, and often rendered more disturbing by attendant physicians applying emergency medicine such as emetics or stomach pumps.

All this changed with Wasson's announcement that he had discovered an indigenous mushroom ceremony believed by many Western experts to be long extinct. Mushroom intoxication was reconceived as a sacred experience with a deep history that long pre-dated Christianity, and Albert Hofmann's isolation of a new psychedelic compound from the Wassons' specimens added the imprimatur of science to the allure of ancient ritual. This was not how Maria Sabina conceived the mushrooms: she herself was a devout Catholic and regarded her ceremony not as an act of worship but simply a healing intervention, 'done with the sole purpose of curing the sicknesses that our people suffer from'.[34] Wasson, however, claimed it as the vestigial survival of a prehistoric mushroom cult that he had long suspected was the source of the world's great religions. 'Magic mushrooms', as the editors of *Life* had christened them, became imbued with qualities that the term 'drugs' was unable to encompass.

Wasson was a curious-minded amateur, but Robert Graves was an eminent authority on ancient religions, and he elaborated their sacred mushroom theories for a wide general readership. The first edition of his popular exploration of classical myth and poetry, *The White Goddess*, published in 1948, made no mention of mushrooms, but in the third edition of 1960 they were everywhere:

a secret Dionysiac mushroom cult was borrowed from the native Pelasgians by the Achaeans of Argos. Dionysius' [*sic*] centaurs, satyrs and maenads, it seems, ritually ate a spotted toadstool called 'fly-cap' (*Amanita muscaria*), which gave them enormous muscular strength, erotic power, delirious visions, and the gift of prophecy. Partakers in the Eleusinian, Orphic and other mysteries may also have known the *Panaeolus papilionaceus*, a small dung-mushroom used still by Portuguese witches, and similar in effect to mescalin.[35]

Graves identified mushroom motifs in Greek and Etruscan art, and determined that Dionysus 'may once have been the Mushroom god'.[36] The Athenian festival of Scirophoria, in which a *skira* or sunshade was carried in the procession, was, he determined, an emblem of the sacred mushroom. The colourful and familiar fly agaric mushroom, he theorised, had been the active ingredient and 'source of prophetic illumination at Eleusis', the ancient Greek mystery rite.[37] Drawing on Siberian accounts of fly agaric mushroom use in shamanism and Nahua (Aztec) records of mushroom use, Graves argued that mushroom use constituted an ancient body of wisdom concealed by priestly elites and later suppressed by Christianity. His theories, often based on poetic associations, acrostics and etymological resemblances, were largely rejected by his fellow classicists but they shaped the psychedelic journeys of the new generation, connecting the experience not just to cutting-edge chemistry and neuroscience but to the mystery religions of deep antiquity.

*　　*　　*

The origin story of psychedelics has long since passed into legend. On 19 April 1943, Albert Hofmann, working in the Sandoz laboratories in Basel, took what he imagined would be a tiny dose of LSD-25, a compound he had synthesised from the ergot fungus five years previously but not sampled at the time. What happened next, as Hofmann bicycled towards his home on the outskirts of the city, has been endlessly replayed and reimagined in psychedelic histories, comic strips and animations, not to mention printed onto blotter squares of LSD itself: the mountain road beneath Hofmann's bicycle tyres warping into a rushing river of rainbow colours, with the surrounding houses and countryside erupting in multi-coloured fireworks, waterfalls and kaleidoscopic spirals. All this is based on the account that Hofmann wrote in his memoir published in 1979, decades after the event, once the tropes of psychedelia were well established. But in his original notes of 1943, submitted three days later to his director of research at Sandoz, Arthur Stoll, the effects are described quite differently. There is no mention of kaleidoscopes and fountains, simply a barrage of distressing psychotic symptoms:

> dizziness, visual disturbance, the faces of those present seemed vividly coloured and grimacing, powerful motor disturbances, alternating with paralysis; my head, body and limbs all felt heavy, as if filled with metal; cramp in the calves, hands cold and without sensation; a metallic taste on the tongue, dry and constricted throat; a feeling of suffocation; confusion alternating with clear recognition of my situation, in which I felt outside myself as a neutral observer as I half-crazily cried or muttered indistinctly.[38]

Hofmann concluded his report to Stoll by describing the experience as comparable to a massive overdose of amphetamines, and speculating that LSD was an unprecedentedly potent drug of this class. In 1962, twenty years later, the story he recounted to the medical journalist Margaret Krieg had become a set-piece anecdote that 'had been recounted many times'; but it was still recognisably the version of 1943:

> I was overcome by the fear that I was going out of my mind, and the worst part of it was, I was clearly aware of my condition; my powers of observation were not impaired. Space and time became more disorganised, but I was not capable of any act of will. I could do nothing to prevent the breakdown of the world around me. At home, the physician was called. I had a feeling as if I were out of my

body. I thought that I had died . . . I even saw my dead body lying on the sofa.[39]

By the time of Hofmann's memoir *LSD, My Problem Child* (1979), however, the psychedelic era was in full swing, and the nightmarish episode of psychosis had assumed the form in which it will be forever remembered:

Kaleidoscopic, fantastic images burst in upon me, alternating, variegated, opening and closing themselves in circles and spirals, exploding in coloured fountains, rearranging and hybridising themselves in constant flux . . . Every sound generated a vividly changing image, with its own particular form and colour.[40]

The other canonical text of the psychedelic movement, from which the term was born, was a quite different case. Aldous Huxley's best-selling essay, *The Doors of Perception* (1954), narrated his first experiment with mescaline, but incorporated ideas that Huxley had already conceived well in advance of the bright May morning in 1953 when he swallowed his famous 400mg of mescaline crystals dissolved in water. Much of the material that he presented as psychedelic epiphany had already appeared in his anthology *The Perennial Philosophy* (1946), which drew together spiritual texts from Christian and Islamic mysticism, and Eastern traditions from the Tao to the Vedanta, aiming to distil their 'Highest Common Factor' into a universal truth. His best-remembered insight into the psychedelic experience, the 'Bergsonian model' of consciousness that conceived the brain as a reducing valve that limits our everyday experience of reality, had already been set out in his introductory letter to Humphry Osmond suggesting the mescaline experiment.[41]

What changed for Huxley was not his interest in mystical experience, which was long established, but his belief that drugs might be a path to it. His conversion was sudden. Well into the 1950s he maintained the view of drugs dramatised in *Brave New World*, that they were dehumanising tools of capitalism and its homogeneous mass culture. In his novels and essays of the early 1950s he wrote extensively about drugs as agents of brainwashing and Pavlovian conditioning, predicting that 'the propagandists of the future will probably be chemists and physiologists'.[42] In the epilogue to *The Devils of Loudun* (1952), the book that preceded *The Doors of Perception*, he lamented that 'millions upon millions of civilised men and women continue to pay their devotions, not to the liberating and transfiguring Spirit, but to alcohol, to hashish, to opium and its derivatives, to the barbiturates, and the other synthetic additions to the age-old catalogue'.[43] Self-transcendence with drugs was, he had argued then, an illusion:

> What seems a liberation is in fact an enslavement . . . For the drug-taker, the moment of spiritual awareness (if it comes at all) gives place very soon to subhuman stupor, frenzy or hallucination, followed by dismal hangovers and, in the long run, by a permanent and fatal impairment of bodily health and mental power . . . This is a descending road and most of those who take it will come to a state of degradation.[44]

Huxley's mescaline experience was irreconcilable with the pejorative notion of 'drugs', just as Graves would reject the term as a description of psilocybin. It was not a drug, he concluded, but rather 'what Catholic theologians call a "gratuitous grace"', not necessary to salvation but, if made available, to be accepted thankfully.[45] In correspondence with Osmond after the event, the pair decided a new word was needed to rescue mescaline and LSD both from the tainted category of 'drugs' and from the clinical terms recently adopted by psychiatrists, 'hallucinogen' and 'psychotomimetic', that linked their effects to the symptoms of schizophrenia. Osmond proposed 'psychedelic', meaning 'mind-manifesting', a framing that privileged positive associations such as mystical insight, mental well-being and spiritual growth. Huxley concurred, though he believed the correct spelling should be 'psychodelic', with which he persisted to little avail.[46]

By this time LSD and mescaline were being widely used in psychiatry and, since their singular effects on consciousness could only be captured by trained human observers, they led to a revival of self-experiment. Mescaline and LSD were quickly joined by new psychedelic compounds, each described in breathless first-person reports by their discoverers. Their style harked back to the scientific self-experimenters of the nineteenth century, and the similarity was more than coincidental. Aldous Huxley drew on this literature, particularly the descriptions of peyote by Havelock Ellis, whom he cited on the first page of *The Doors of Perception*. He interrogated his experience through art-historical comparisons just as Ellis had done, and on occasion alighted on the same touchstones. The poetry of William Wordsworth, for instance, who Ellis predicted would be 'the favourite poet of the mescal-drinker',[47] was for Huxley 'a new direct insight into the very Nature of Things'.[48]

Ten years previously, Huxley's *Perennial Philosophy* had been reviewed by literary critics as an example, in Humphry Osmond's paraphrase, of the 'unfortunate mystical trends in his later years'.[49] By the mid-1950s, and with the sheen of modern science that mescaline lent them, these ideas were in tune with the new zeitgeist. The reaction against the behaviourist paradigm was now a broad church. The Freudian renegade Wilhelm Reich located neurosis in the strictures of a rigid mass society and proposed a sexual revolution as the remedy. Herbert Marcuse's critique of the 'one-dimensional man' created by the 'affluent society' of consumerism fed a desire for authentic personal experience that transcended mass culture. Carl Jung's theory of individuation proposed a lifetime of personal growth, spurred by dreams and spiritual practice, through which the subject could carve out their unique path to wholeness and fulfilment.

The rediscovery of introspection and mystical experiences revived interest in the work of William James, particularly *The Varieties of*

Religious Experience. This was central to the new 'humanistic psychology' of the US professor Abraham Maslow, who introduced many of the concepts that would frame the new understanding of psychedelics: the hierarchy of needs, self-actualisation and the 'peak experience'. Maslow found mid-century psychology mechanistic and reductive, and distrusted its claim to be objective and value-free, operating as it did as the tool of a regimented and authoritarian society. Alongside William James, he drew inspiration from an eclectic range of teachers that included the Freudian Alfred Adler, the anthropologist Ruth Benedict and the founder of Gestalt psychology Max Wertheimer. Beginning with 'A Theory of Human Motivation' (1943), his work moved beyond psychiatry's traditional focus on the sick mind to establish the bases for mental health, happiness and fulfilment.

In Maslow's definition, the human was 'a permanently wanting animal', who was 'most often partially satisfied and partially unsatis-fied', with 'increasing percentages of non-satisfaction as we go up the hierarchy' from basic needs such as food and shelter towards creative and spiritual fulfilment.[50] For a post-war generation whose material requirements were substantially met, the most pressing needs were social satisfactions such as self-esteem and the respect of others and, at the top of the hierarchy, 'self-actualisation', in which the individual has fulfilled their potential on every level. The summit of self-actualisation was the 'peak experience', an ecstatic moment in which the individual achieves total self-awareness and harmony with their surroundings. This was a state that transcended social structures: it was democratic and ecumenical, with no religious belief required.

Maslow's major book of the 1960s, *Religions, Values and Peak Experiences* (1964), grafted the psychedelic experience onto his existing theories just as Robert Graves's new edition of *The White Goddess* had done. Building on the ideas of Aldous Huxley and Alan Watts, he claimed Huxley's psychedelic 'self-transcendence' as a paradigm for the 'peak experience', which in turn became a paradigm through which psychedelic enthusiasts articulated their experiences. Psychedelics

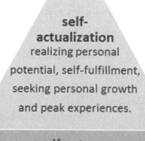

Abraham Maslow's Hierarchy of Needs

generated peak experiences; peak experiences helped people to self-actualise, and self-actualisation made peak experiences more attainable. Maslow never took psychedelics himself on account of a heart condition, but also because he felt there was no need: he had read William James, and had 'been there' already.[51]

By the time that *The Doors of Perception* became a bestseller, Sandoz Pharmaceuticals had made LSD available to psychiatrists. Their internal trials of the drug had left them unsure of what diagnoses it should be indicated for, and in what doses; their solution was to offer it free to psychiatrists on an experimental basis, in return for their clinical recommendations. Among those who took up Sandoz's offer was the Los Angeles psychotherapist Oscar Janiger, whose celebrity clients publicised LSD to their mass audience. In 1959, before Timothy Leary had taken his first mushroom trip, Cary Grant stood on the set of his film-in-progress *Operation Petticoat* and announced to the press – and to the horror of his studio publicists – that LSD therapy with Dr Janiger had changed his life:

> It was absolute release . . . We come into this world with nothing on our tape. We are computers, after all. The content of that tape is supplied by our mothers, mainly because our fathers are off hunting or shooting or working. Now the mother can teach only what she knows, and many of these patterns of behaviour are not good, but they're still passed on to the child. I came to the conclusion that I had to be reborn, to wipe clean the tape.[52]

In *The Varieties of Religious Experience*, William James had proposed a distinction between two character types that led to 'two different conceptions of the universe of our experience'. One type he called the 'once-born': those who accept their lives as given and attempt to make the best of them. For the other class, the 'twice-born', the task was different: 'peace cannot be reached by the simple addition of pluses and elimination of minuses from life'. The twice-born needed to begin afresh; they were those, in James's scheme, who experienced sudden religious conversion or transformative existential insights along the lines of Cary Grant's. Although James referred to the once-born as 'healthy-minded' and the twice-born as 'sick souls', the thrust of his distinction was to undercut moral judgement: one mindset, as he insisted, cannot be criticised from the standpoint of the other. He also suggested that the once-born mind, healthy as it may be, is inclined to smug and dull conservatism, while the twice-born seeks adventure and transformation.[53] Psychedelics were agents of radical change for the twice-born, and of peak experience for all.

* * *

During the watershed years of the early 1960s, the stigma attached to madness and abnormal mental states was challenged on multiple levels. The blows landed with stunning rapidity. In 1961 the sociologist Erving Goffman published *Asylums*, a study of one of the vast mental hospitals that had mushroomed during the post-war years, exposing them as machines for 'institutionalisation' that eroded their patients' life skills and identities. The same year *The Myth of Mental Illness*, by the New York psychiatrist Thomas Szasz, dismissed psychiatry as a modern priestcraft that pronounced diagnoses of mental illness much as the churches had once persecuted heretics or witches. Michel Foucault's monumental *Histoire de la folie à l'âge classique*, first published in 1961 and later translated into English in abridged form as *Madness and Civilization*, argued that the Age of Reason had forged a version of modernity in which the voices of the mad – previously a polyphony of nonsense, allegory, wit and wisdom – could no longer be understood, only pathologised and punished. In 1960 the Scottish psychiatrist R.D. Laing published *The Divided Self*, his radical rethinking of schizophrenia that aimed to rescue it from psychiatry and understand it as a crisis with the potential to heal a mind tormented by insoluble problems. Laing in particular would steer a course that intersected with the drug culture of the 1960s, giving LSD to his patients and claiming it as a tool of mental liberation that broke down the barriers between madness and sanity, to the benefit of both.

In 1962 *One Flew over the Cuckoo's Nest*, the bestselling novel by the young Ken Kesey, dramatised the insights of Goffman, Szasz, Foucault and Laing for a mass readership, turning them into a parable of a society that enforced conformity by crushing mental alterity. Kesey based the institutional setting on his experiences in the Veterans' Hospital at Menlo Park in Stanford, where he had volunteered for a trial of psychedelic drugs including LSD, psilocybin and mescaline.[54] The experiments were directed by a psychopharmacologist, Leo

Hollister, who had become sceptical about the prevailing assumption that psychedelics produced a 'model psychosis' that duplicated the symptoms of mental illness. Hollister launched a study in which the psychotic symptoms of mental hospital patients were compared with drug-induced 'psychoses' in healthy volunteer subjects, the role for which Kesey was recruited with a starting fee of $25 per trial session.

Hollister's findings, published in 1962, exposed the psychotomimetic model of psychedelics to scrutiny from which it never recovered. The central problem, he concluded, was the term 'hallucination', which sounded like an accurate clinical term but in practice was used to describe dozens of often unrelated phenomena. Hollister showed that psychedelics rarely induced true hallucinations, let alone 'mental syndromes comparable to schizophrenic reactions'.[55] Most typical symptoms of schizophrenia such as 'disorientation, paranoid ideation, disturbed thinking, and auditory, gustatory, olfactory or tactile hallucinations' were uncommon on psychedelics, particularly with healthy subjects taking them in relaxed and benign environments.[56] Patients with diagnoses of schizophrenia were easily able to distinguish the transient effects of LSD from their long-term condition.

The psychotomimetic model had driven clinical research into psychedelics for a decade, and had been adopted by many of its early advocates. Aldous Huxley opened *The Doors of Perception* with the claim that mescaline's effects 'are similar to those which occur in that most characteristic plague of the twentieth century, schizophrenia';[57] Robert Graves described his psilocybin trip as 'a perfect schizophrenia'.[58] But now the science was moving on. Emergency hospital wards were seeing an increase in patients presenting with classic symptoms of psychosis, such as voice-hearing and delusions of persecution, who turned out to have taken large doses of amphetamines. The term 'amphetamine psychosis' was coined in 1958, and clinical research into its biological basis shifted to the role of the dopamine system, which underpinned the action of both amphetamines and the new antipsychotic medications. Psychedelics were no longer pathologised as a chemical form of psychosis, but at the same time the dazzling prospects they had offered for curing schizophrenia faded away.

*   *   *

The Harvard Psilocybin Project offered Timothy Leary and Richard Alpert an opportunity to convene all the early adopters of psychedelics, from beatniks to mystics to scientists, and to map a consensus for further exploration. They made plans for them all to take Indocybin together and summoned each in turn, but the exercise only highlighted their differences. Aldous Huxley, struggling with his utopian psychedelic novel *Island*, was delighted to learn that the new drugs were to be studied at Harvard, reviving the legacy of William James, and he arrived with his fellow mystic Gerald Heard and Humphry Osmond. Allen Ginsberg's appearance disrupted the scholarly mood, and his antic recitals of drug-inspired poetry left the psychiatrists and Harvard faculty members baffled. He introduced Leary to cannabis and, eager to spread the word among his network of anarchists and sexual revolutionaries, took a stash of psilocybin pills back to New York, where he circulated them at the Five Spot café in the Bowery to regulars including the jazz pianist Thelonious Monk.

Richard Alpert insisted on inviting William Burroughs, who 'knows more about drugs than anyone else alive', but his arrival only made things worse.[59] Burroughs's pioneering and globetrotting experiences with psychedelics – peyote in Mexico, ayahuasca in the Amazon, DMT in London and Tangier – had inspired him with visions but also left him shell shocked and traumatised, and he regarded their enthusiastic adoption by psychiatry and the CIA as a first step towards their use as weapons in the coming era of mind control. He sat in his fedora hat, drinking gin and tonics and regarding the scene with scorn. He quickly tired of Leary's revolutionary boosterism and left. 'Their Immortality Cosmic Consciousness and Love is second-run grade-b shit,' he wrote later in *Nova Express* (1964). 'Flush their drug kicks down the drain.'[60]

Leary was unable to unite these disparate perspectives, still less integrate them into a university research project that had devolved into an ongoing twenty-four-hour party. David McClelland, the professor who had brought Leary in to invigorate the Harvard Center for Personality Research, was disappointed that the proposed drug trials had been subsumed into a freewheeling lifestyle experiment: 'One can hardly fail to infer that one effect of the drug is to decrease responsibility or increase impulsivity.'[61] Huxley became exasperated by Leary's 'nonsense-talking', which he saw as 'just another device for annoying people in authority, flouting convention, cocking snooks at the academic world'. 'I am very fond of Tim', he confessed to Osmond, 'but why, oh why does he have to be such an ass?'[62] A scathing editorial in the *Harvard Crimson* depicted the project as a performative rebellion against the academy:

> The shoddiness of their work as scientists is the result less of incompetence than of a conscious rejection of scientific ways of looking at things. Leary and Alpert fancy themselves prophets of a psychic revolution designed to free western man from the limitations of consciousness.[63]

In March 1962, the Food and Drug Administration (FDA) was alerted to the proceedings, and ruled that a medical doctor must be present whenever the drug was administered. At this point Leary decided that the scientific 'game' was played out, and psychedelics were more important than Harvard.

The FDA's close attention to the Psilocybin Project was an early indication of developments that would impact the future of psychedelic science much more profoundly than Leary's dismissal from the university. In October 1962 President John F. Kennedy signed into law a series of new measures, known as the Kefauver-Harris Amendment, that gave the Federal Food, Drug and Cosmetic Act more powers to ensure that pharmaceutical drugs were safe and effective. Under the amendments, the process of approval for new drugs became more stringent. Their efficacy needed to be demonstrated by randomised control trials against a placebo, in clinical conditions that excluded any 'extra-pharmacological variables'. All adverse reactions were to be reported to the FDA, with more data demanded before proceeding to human trials. The therapeutic application for a new drug was to be specified in advance of its trial, and the benchmark of quality in a pharmaceutical product was how accurately its effects could be reproduced.

The Kefauver-Harris Amendment was not directed at psychedelic research. Specifically, it was a response to the tragedy of thalidomide, which had not been approved by the FDA and had led to thousands of children being born with birth defects. Its broader targets were similar to those of the Pure Food and Drug Act of 1906: inaccurate advertising, quack therapies, the poisoning of consumers by large corporations and the testing of new drugs on captive human populations such as patients or prisoners. Nonetheless, its effect on the psychedelic field was shattering. Given the huge variation in their effects on different subjects, LSD, mescaline and psilocybin were the very opposite of 'quality' drugs under the new definition. The effects they produced could not be measured by any of the conventional biomarkers. Placebo trials were bound to be flawed, since a pill with no psychoactive effects was immediately identifiable as such. 'Extra-pharmacological variables' were impossible to exclude, since the effects of psychedelics were so profoundly determined by mood and context – as they were now known, 'set and setting'. Human trials were the only type of research that could demonstrate their effects, and specific applications could not be determined in advance.

The amendments reflected new approaches to medical risk, which it aimed to control by steering drug development towards measurable benefits and precise disease indications. This worked well for some types of drug, anti-bacterials for example, where biomarkers and pathogens were easily identified and competing treatments directly comparable, either against placebo or one another. But its effect on mental treatments was perverse and often counterproductive. Its implicit assumption, that all drugs worked through direct biochemical action, encouraged over-rigid diagnostic categories for mental illness, paired with overspecific 'magic bullet' medications to address them. The exclusion of qualitative data in favour of the largest possible cohort trials effectively marked the end of medical self-experiment. Human trials could no longer be undertaken speculatively, without a pre-established medical proposition to test; nor could subjective testimony inform the process, since quality was now determined by the overall or average outcome. Descriptions of subjective experience were relegated from data to anecdote.

There was no explicit ban on psychedelic drug research – LSD trials continued at the Spring Grove Clinic in Maryland, for example, well into the 1970s – but the new FDA regime had a chilling effect. It became harder to fund trials, and more onerous to approve drug applications. Psychedelic trial designs, in attempting to comply with the new amendments, produced incompatible results. Some chose to focus on the physiological action of the drug, others on the psychotherapy that accompanied it. Some took place in hospitals, others in community settings. With extra-pharmacological variables rigorously excluded, the overall picture was confused: some trials suggested that smaller doses were more effective, others larger. Some elicited a mix of strikingly positive and negative outcomes, which were smoothed out into unremarkable average scores that represented neither. By the end of the decade, psychedelic science had lost its fashionable appeal in academia and funding applications were drying up.

As so often during these Janus-facing years, however, as one door closed another opened. In April 1960 the thirty-five-year-old organic chemist Alexander Shulgin, his curiosity piqued by reading *The Doors of Perception*, was administered Huxley's dose of 400mg of mescaline by a psychologist friend at his home outside Berkeley. The experience, Shulgin wrote later, 'unquestionably confirmed the entire direction of my life'. He was transported by colours that seemed to extend beyond the visible spectrum, and by the exquisite detail of the living world –

'the intimate structure of a bee putting something into a sac on its hind leg to take to a hive' – but his central insight was the one that his predecessors, from Havelock Ellis to Aldous Huxley and Robert Graves, had found in the poetry of William Wordsworth:

> More than anything, the world amazed me, in that I saw it as I had when I was a child. I had forgotten the beauty and the magic and the knowingness of it and me. I was in familiar territory, a space wherein I had once roamed as an immortal explorer, and I was recalling everything that had been authentically known to me then, and which I had abandoned, then forgotten, with the coming of age.[64]

Shulgin worked as an industrial chemist, mostly in his home laboratory, where he also collaborated with the Drug Enforcement Agency on identifying novel street drug compounds. Over the next few years he synthesised dozens of new psychoactive drugs, mostly phenethylamines related to mescaline. These included the N-methylated version of Gordon Alles's amphetamine variant, MDA, known as MDMA, which spread rapidly through the underground and dance cultures of the 1970s under the street name 'ecstasy'.[65] Shulgin was a principled self-experimenter who believed that the subjective encounter with a new drug was an integral part of the process of discovery. The chemical structure of a molecule, he observed, gives no clear indication of whether it will be psychoactive, or what its effects will be:

> These properties cannot yet be known, for at this stage they do not yet exist ... It is only with the development of a relationship between the thing tested and the tester himself that this aspect of character will emerge, and the tester is as much a contributor to the final definition of the drug's action as the drug itself.[66]

Since he had no interest in patenting his compounds or getting them approved for medical prescription, Shulgin was able to work

32. The chemist Alexander Shulgin synthesised dozens of new psychedelic compounds and devised a self-experimental protocol to assess their effects.

outside the FDA's new guidelines. In their place, he and his associates instituted a self-experimental protocol of their own devising. The rules for dosage were to begin with between 10 and 50 times less by weight than the expected active dose, based on the closest analogue compounds – a more cautious version of Aleister Crowley's 'Chancery Lane rule' – and to double that dose on alternate days until an effect was attained. At this stage the Shulgin Rating Scale was applied, a simple self-assessment of the drug's intensity. 'Minus' meant no effect, 'plus-minus' a non-specific threshold effect, and 'plus-one' a clear perception that the drug was working, even if the effects were only physical responses such as nausea. 'Plus-two' denoted a marked perceptual or sensory effect, capable of detailed description in visual, tactile or emotional terms, usually accompanied by cognitive impairment of the kind that would make it undesirable to take a phone call or unwise to drive a car. 'Plus-three' was the drug at its maximum intensity. Beyond that was 'a serene and magical state which is largely independent of what drug is used': this was not an intensification of plus-three but its own category, for which Shulgin used Abraham Maslow's term 'peak experience'.[67]

Shulgin went on to publish hundreds of drug syntheses in two door-stopping volumes, *Pihkal* (1991) and *Tihkal* (1997), that described each compound first by chemical synthesis and then by 'subjective extensions and commentary', culled from the notes and descriptions of his own experience and those of his volunteer subjects. The format returned psychedelic science neatly to its origins in Humphry Davy's pioneering study of nitrous oxide, *Researches Chemical and Philosophical . . .* (1800), which opened with his laboratory experiments and concluded with the reports of his human subjects – doctors, poets and writers – on their subjective responses to a lungful of the gas. As Davy and Shulgin both recognised, the languages of pharmacy and introspection may be mutually incomprehensible, but a full account of a drug and its psychoactive effects requires them both.[68] Shulgin's volumes became the bible for a loose cohort of underground chemists and researchers working outside the academies, universities and laboratories of the scientific establishment, often anonymously and using the self-descriptive label of 'psychonauts'.

* * *

By the end of 1962 – before the Beatles' first LP, let alone the first Acid Test parties and Magic Bus trips – the script for drugs in the twenty-first century had essentially been written. They emerged from those watershed years delicately balanced on a fault-line between two versions of modernity. Under international law and in the domain of institutional science, the strictures of the Progressive Era were cemented and extended: drugs were globally criminalised, and their scientific and medical uses strictly controlled. At the same time, the seeds had been planted for their revaluation by a coming generation of intellectually curious global consumers, with an ethos of heroic self-experiment and a gaze fixed on the horizon of self-transcendence.

AFTER DRUGS

In the twenty-first century, the monolithic category of 'drugs' has fissured and blurred. Many of the substances that were prohibited in the twentieth century are no longer confined to street dealers and the criminal underworld. Cannabis can now be bought in sleekly designed high street stores across the USA and Canada; in Europe, the cannabis 'coffee shops' of the Netherlands are poised to spread across the continent, along with the Spanish model of membership-based 'cannabis clubs', producing for their own consumption and tolerated by local law enforcement. Prescription sedatives and stimulants are available from unlicensed online suppliers around the globe, and every mind-altering substance imaginable can be found on the marketplaces of the Dark Web. In Silicon Valley, a goldrush is under way to secure patents and licences for psychedelic drugs such as psilocybin and MDMA, as they advance painstakingly through the FDA trials that promise to legitimise their use in clinical psychotherapy.

The Progressive Era view of drugs as a social problem has by no means disappeared, but alongside it their expansive possibilities that were explored in the nineteenth century are being rediscovered. The modern search for mental stimulants has created a galaxy of cognitive enhancers and nootropics, combining 'brain-booster' chemical supplements with ayurvedic herbs, Chinese mushrooms, 'activated' roots and colloidal silver, in a market valued in 2018 at over $2 billion and set to double by 2025.[1] Like the electropathic belts and gland therapies of the nineteenth century, their benefits and dangers are hotly debated, and their futuristic possibilities inform the fictional descendants of *Dr Jekyll and Mr Hyde* and *The Strand Magazine* stories of self-experiments gone wrong: novels and films such as *Limitless* (book 2001, film 2011, TV series 2015–16) and the plethora of streaming-TV series in which designer drugs or pharmaceutical trials precipitate the protagonists into strange metamorphoses, psychotic breaks or alternate realities.[2]

Despite the taboo on self-experiment within academic science, the last decade has seen a resurgence in the use of drugs for consciousness exploration. Today's researchers typically deploy powerful short-acting psychedelics such as DMT or ketamine rather than volatile anaesthetics such as ether or chloroform, but their accounts of disembodied existence range, similarly to those of the nineteenth century, from contact with non-human entities to transcendental mystical experiences, survival after death to hidden dimensions of the mind.[3] Drugs are equally prevalent in the modern subcultures of spirituality, magic and the creative imagination, where practices drawn from shamanism, Wicca and paganism blend with the immersive fantasy worlds of fiction, digital adventures and role-playing games. The twenty-first-century term 'occulture', in which art, literature and magical practices are linked as they were for the symbolists and spiritualists of the *fin-de-siècle*, marks out a territory steeped in drug experiences, thanks in no small part to the looming posthumous presence of Aleister Crowley.

Today, altered consciousness and non-ordinary reality are primarily associated with the psychedelics that emerged in the 1950s but, as we have seen, explorations of this kind date back much further: psychedelics have colonised a pre-existing cultural niche that flourished long before the term was coined. During the nineteenth century, when the drugs we now class as psychedelic were present only at the margins, inner voyages of this kind were undertaken with an eclectic mix of substances; some of them, such as hashish or nitrous oxide, have found their own niche in the modern drugs landscape, while others, such as ether or chloroform, have been largely forgotten. The category of 'psychedelic', in addition to its various pharmacological definitions, has come to include the sense that these particular compounds offer a unique conduit to spiritual experience, mental healing or a higher reality.[4] Just as the term 'drugs' was loaded from the beginning with the pejorative associations embedded in the Progressive Era's version of modernity, 'psychedelics' bears the imprint of the rebirth of inner experience half a century later. Both terms carry a concealed set of value judgements, modern analogues of the profane and the sacred.

The twenty-first-century embrace of psychedelics has been hailed as the beginning of the end of the 'War on Drugs', but it has also reinforced and entrenched the category of 'drugs' by excluding less heavily stigmatised substances from it. This reframing of the problem has been described, notably by the Columbia University professor of psychology Carl Hart, as 'psychedelic exceptionalism'. For Hart, writing from a Black perspective, it is no coincidence that the substances in this category are the drugs of choice for white, college-educated users: he sees their advocates as 'strategically protecting their mission to ensure public support for a select few psychedelics' by separating them from the wider, racially inflected 'drug problem'.[5] The colour line identified by W.E.B. Du Bois still runs conspicuously through the drug laws and their enforcement. Behind the boutique cannabis stores and Silicon Valley psychedelic start-ups the War on Drugs grinds on, with its conveyor-belt drug courts, asset forfeiture, mandatory drug testing and custodial sentences. The number of drug arrests in the US today is virtually unchanged since the late-twentieth-century peak of the War on Drugs.[6] Drug speech remains heavily censored: images of drugs are excised from social media platforms, and many internet service providers block access to websites with drugs content (including drug information and health services). The Progressive Era view of 'drugs' remains institutionally entrenched, and the war against them is still disproportionately targeted at ethnic minorities and marginalised social groups.

Another feature that Hart notes in today's psychedelic discourse is its silence on the question of pleasure. He recalls a 'middle-aged white military veteran' accosting him in the Columbia gym to share his appreciation of psychedelics, which he referred to as 'plant medicines'. 'It was particularly important to him that I know he "didn't get high"', Hart wrote, 'and only used the plants to facilitate his "spiritual journey".' Hart stumped him with his deadpan response: 'What's wrong with getting high?'[7] The nineteenth-century debates over euphoria, in which drug-induced pleasure became suspect or pathological, have been succeeded by a twenty-first-century neuroscience in which positive mood is seen as a symptom of raised dopamine or serotonin levels, and interpreted through the language of brain reward mechanisms, craving and addiction. Clinicians' fear of encouraging drug-seeking behaviour makes pleasure into a negative symptom: for example, esketamine, the ketamine analogue licensed as an antidepressant by the FDA in 2019 and marketed by Janssen Pharmaceuticals as Spravato®, includes at the head of its long list of common side-effects: 'feeling extremely happy ("euphoria")'. Hart – like Moreau, Mantegazza and Freud – considers pleasure as a positive, with the potential to trigger a wide range of physical, mental and social benefits. 'Pleasure is a good thing, something that should be embraced,' he insists, adding: 'It feels weird that I am compelled to write the preceding sentence because the idea seems so obvious.'[8] The advocates of psychedelics prefer to describe them as 'medicines', 'entheogens' or 'sacraments': terms that avoid the stigmatised label of 'drugs', and also sidestep the possibility of pleasure as a motive for their use.

Hart has been a stern critic of institutional drug science and particularly of the US federal agency, the National Institute of Drug Abuse (NIDA), which funded many of his studies and on whose advisory board he served. As its name suggests, NIDA has focused almost entirely on funding research that identifies drug-related harms and presents them as emphatically as the data allows. Career advancement, for Hart as for others, depended on internalising the belief that drugs were a danger from which the public must be protected: a belief he himself subscribed to for many years despite clear evidence that the drug-using subjects with whom he worked were deriving positive outcomes from their use. Only after a decade of such studies did the evidence force him to conclude that all drugs, not just psychedelics, had the ability to make their subject feel 'more altruistic, empathetic, euphoric, focused, grateful, and tranquil'.[9]

*   *   *

The erasure of self-experiment in the 1960s was followed by a series of neuroscientific discoveries that placed consciousness-altering drugs at the centre of a new understanding of the brain. In the 1970s, the neuro-chemical receptors that modulated the effects of opiates were identi-fied, and morphine-like peptides isolated from the pituitary gland were named 'endorphins', a contraction of 'endogenous morphine'. Opiates, it turned out, worked by mimicking the chemicals used by the brain to dull pain and prompt euphoria in response to food, sex and drugs. Psychedelic drugs were subsequently shown to function by boosting levels of serotonin, a neurotransmitter originally discovered in 1948; stimulants such as cocaine and amphetamine flooded the brain with dopamine. The powerful dissociative effects of ketamine and phency-clidine (PCP) were found in 1983 to be caused by blocking glutamate, the brain's most abundant neurotransmitter. In 1992 the Israeli chemist Raphael Mechoulam isolated the first endocannabinoid, and demon-strated that it bound to the same receptors as THC, the psychoactive ingredient in cannabis. As the brain gave up its neurochemical secrets, it revealed itself as a factory of mind-altering – and in many cases illegal – drugs.

The revolution in neurochemistry brought intense scientific scru-tiny to bear on the effects of these drugs, but the new methods were very different from those of the nineteenth century. With their massively expanded toolkit, neuroscientists are able to study drug actions at one remove, by examining their physiological correlates in brain scans, and their average effects via data from randomised control trials. Institutional research, such as that funded by NIDA, focuses not on subjective experience but on the brain itself, identifying receptor sites and reward mechanisms that might offer clues to the treatment of drug addiction. A popular-science jargon has emerged that recycles crude characterisations of neurotransmitter activity: 'serotonin' has become a shorthand for happiness, and 'dopamine' for the cycle of craving and reward.

The 'Decade of the Brain', as the 1990s was designated by President George H.W. Bush, proved to be the springboard for a revival of scientific interest in drugs, and particularly the human study of psychedelics that had foundered after the FDA amendment of 1962. In 1990 the clinical psychopharmacologist Rick Strassman obtained US government funding for trials of the potent and short-acting psychedelic DMT at the University of New Mexico, in which his subjects recounted near-death and mystical experiences. By this time, the drug activists Rick and Sylvia Doblin had founded the Multidisciplinary Association for Psychedelic Studies (MAPS) with the aim of integrating psychedelics into clinical psychotherapy. In 2008 MAPS completed its independently funded pilot study of MDMA (ecstasy) in the treatment of post-traumatic stress in military veterans, and in 2017 it was granted 'breakthrough therapy status' by the FDA, allowing it to progress to clinical trials that would allow MDMA to be licensed as a prescription medicine.

Over the last decade the trickle of psychedelic research has become a flood. Strassman's study opened the path for universities, from Johns Hopkins in Baltimore to Imperial College in London to the University of California at Berkeley, to establish laboratories and departments to study the effect of psychedelic drugs on the brain. MAPS is now at the centre of a constellation of research institutes, venture capitalists, pharmaceutical corporations and Silicon Valley entrepreneurs developing psychedelics for use in clinical therapies. The exponential growth of psychedelic science in the twenty-first century has put the prohibition on self-experiment and subjective reportage under strain. Many of the leading researchers in the field were drawn to it by their own psychedelic experiences, and continue to self-experiment informally. Others maintain principled objections: subjective experience runs the risk of skewing or distorting the clinical data, or introducing biases towards solipsistic and non-measurable findings. Even if these risks are avoided, self-experiment can negatively impact confidence in the project's integrity.[10] The case in favour is rarely argued explicitly, but was articulated in 2007 by Felix Hasler of the Berlin School of Mind and Brain at Humboldt University:

> There are two classical positions. Some people say that one shouldn't do self-experiments because this jeopardises scientific objectivity. I don't agree with that. If I do hallucinogen research, I should know the effects of these substances first-hand. Besides, there is an ethical responsibility. If I expect my test subjects to put up with certain states, I should at least know from personal experience what they're going through.[11]

Hasler's arguments echo through the long history of self-experiment: from the principled position of J.B.S. Haldane to the practical psycho-pharmacology of Alexander Shulgin, and as far back as the Royal Society's 'Nullius in Verba'. Today it exists at the fringes of a scientific paradigm constructed from objective and measurable data, where subjective experience struggles to find purchase.

On the borders of neuroscience and psychiatry, however, self-experiment has found some productive niches. The German neurophilosopher Thomas Metzinger, for instance, draws on his experiences with LSD, DMT and mescaline to inform his theories of subjectivity and the self; the British psychologist David Luke immerses himself in the peyote rites of the Huichol people of northern Mexico to study the role played by the cactus in their practices of clairvoyance and healing. The prohibition on self-experiment can itself be powerfully deployed. In his influential contribution to gender studies, *Testo Junkie* (2008, originally writing as Beatriz Preciado), Paul Preciado used self-experiments with testosterone to explore the history of sexuality and its medical hinterland. Self-describing as a 'gender pirate' or 'gender hacker', he experimented outside the parameters of licensed medicine:[12]

> One's relationship to testosterone changes as soon as one leaves the framework of medical and legal protocol for changing sex . . . outside the institutional context defined by the state, testosterone is no longer part of a therapy of hormonal substitution and becomes an illegal drug, just like cocaine or heroin.[13]

Preciado cites Sigmund Freud's cocaine experiments as a precedent for 'the absorption of new technologies of the modification of subjectivity', and claims the Steinach operation to which Freud submitted himself in his later years as a precursor to his own hormone experiments.[14] He sees self-experiment – 'the principle of the auto-guinea pig' – as democratic and empowering, 'a mode of the production of "common" knowledge and political transformation' that needs to be reclaimed from institutional medicine and directed towards 'the coming liberation movements of gender, sexual, racial, and somatic-political minorities'.[15]

*   *   *

The world from which the term 'drugs' emerged is receding from view. The social solidarity of the Progressive Era has given way to an atomised individualism in which policing the private consumer choices of millions is no longer possible. When Richard Nixon relaunched the 'War on Drugs' in 1971, cannabis (or 'marijuana') was an alien commodity to most people over the age of thirty. Today, most people below retirement age recognise drugs, for better or worse, as part of modern culture. We are global consumers, at home with the novel and exotic in everything from food to music, travel to spirituality; our appetite for intoxicants participates in this pursuit of novelty and sensation, and is linked to it by consumer advertising that borrows from the iconography of mind-altering drugs to sell us everything from energy drinks to holidays to smartphones. 'Drugs', when the term appeared around 1900, drew on a reflexive distrust of the alien that has itself become alien to the inhabitants of the twenty-first century.

In the interim, it has vaulted over the original, neutral sense of the word 'drug' to become its primary meaning. Although the broader sense of the word is retained in some everyday usages, it is slowly receding: medical drugs, for example, are now more commonly referred to as 'medications' or 'meds' to avoid confusion or stigma. The category of 'drugs' remains embedded in our laws for the foreseeable future, but it is tempting to imagine what might lie beyond it. If it were to disappear, there would be no need to replace it. A post-drug world would need not a new language, but the recovery of an older one. Behind it is a disparate group of plants and chemicals whose properties offer more meaningful terms: stimulant, sedative, narcotic, psychedelic, euphoriant. If we needed a portmanteau term, we could use the most obvious and neutral one: 'psychoactive', or mind-altering. Value judgements, both eulogising and demonising, are already abundant, and new ones will doubtless emerge.

The invention of 'drugs' can be seen as an attempt, characteristic of its historical moment, to draw a sharp line between good and bad substances and enforce it with state control. Yet we have known since antiquity that good and bad are not inherent in any particular plant or molecule. As the classical authority Dioscorides wrote over two thousand years ago, 'the dose makes the poison': all drugs have the potential to heal or to harm. We are, in Sigmund Freud's phrase, prosthetic gods; the meaning of drugs lies not in the substances but in our selves.

ENDNOTES

PROLOGUE BEFORE DRUGS

1. Davis 1998 214
2. Most 1984
3. Davis and Weil 1992 56
4. Davis 1998 235
5. 3 June 1994
6. Davis 1998 235
7. Hart 2021 93
8. For example, 'Highway Highs: Keeping Tabs on the Net is Harder than Learning How to Make Them', *Arena* 53, September/October 1995, 50–2.
9. Digitized examples of this pre-internet subculture can be found at the indispensable www. erowid.org. In the UK, the archive of DrugScope, a drug workers' charity that assembled an extensive library of underground drug user publications from this period, is now housed at the Wellcome Collection.
10. Baudelaire 1996 [1860] 71. The two best-known studies at the time were Alethea Hayter's *Opium and the Romantic Imagination* (1968) and Molly Lefebure's *Samuel Taylor Coleridge: A Bondage of Opium* (1974), both now classics in what has since become a crowded field.
11. Hofmann 2013 113
12. Arnold 1869 vii
13. Perry 1948 363
14. Letter from Jones to James Strachey, Freud's English translator, in Ferris 1997 59
15. Uthaug et al. 2019
16. Biden 2021 205
17. Sherwood et al. 2020
18. Pollan 2018 277
19. Taylor 1989 3
20. Ibid. 393

I THE ELIXIR OF LIFE

1. 21 April 1884, in Byck (ed.) 1974 6
2. Freud 1884, in Byck (ed.) 1974 53
3. Freud, letter to Wilhelm Fliess, 1 Feb. 1900, in Masson (ed.) 1985 397–8
4. Oppenheim 1991 84
5. Beard 1881 96
6. Ibid. 103
7. Ibid. 98

8. Ibid. 99
9. Ibid. 100
10. Freud 1884, in Byck (ed.) 1974 65
11. Ibid. 64
12. Balzac 2018 [1829] 18
13. Ibid. 23
14. Ibid. 24
15. Ibid. 25
16. To Michael Foster, 18 September 1844, *Medical History Supplement* 28 (2009), 105–33
17. Brown-Séquard 1889 105
18. Ibid. 106
19. Phillips 2014 73
20. Aschenbrandt 1883 732
21. Bentley 1880, in Byck (ed.) 1974 16
22. von Bibra 1995 [1855] 85
23. Ibid. 87
24. Ibid. 92
25. A 'drachm', or dram, is one-sixteenth of an ounce, or a little over 3.4 grams. The avoirdupois system of measurement used by apothecaries and early pharmacists began with a 'grain' (roughly 65mg); 20 grains made a 'scruple' (around 1.2 grams), three scruples a drachm, and 8 drachms an ounce.
26. Mantegazza 1859, in Andrews and Solomon (eds) 1975 39
27. Ibid. 40
28. Ibid. 42
29. Samorini 1995 17
30. Ibid. 18
31. Ibid. 20
32. Ibid. 16
33. Freud 1884, in Byck (ed.) 1974 58
34. Ibid.
35. Ibid. 60
36. von Bibra 1995 [1855] 92
37. In Byck (ed.) 1974 98–9
38. Newton, *Of Colours*, MS Add. 3975, pp. 1–22, Cambridge University Library
39. Hooke records being 'mightily refreshed by tobacco' in his diary of 17 July 1678 (London Metropolitan Archives, CLC/495/MS01758)
40. Hooke 1726 210
41. 1 October 1672, in Hooke 1935
42. Breen 2019 101
43. Schaffer 1992 339
44. Franklin 1837 [1785] 56
45. Ibid. 18
46. 18 April 1799, Davy to Davies Giddy, http://www.davy-letters.org.uk. In 'The Atmosphere of Heaven' (Jay 2009), I tell the full story of the Pneumatic Institution and the nitrous oxide experiments.
47. Davy 1800 458–9
48. Stansfield 1984 166
49. Davy 1800 496
50. Ibid. 496
51. Ibid. 488–9. For further discussion of Davy's epiphany and its sources, see Jay 2009 198–9.
52. Strickland 1998 458–9

53. Goethe had been wary of coffee's stimulant powers in his youth, when his fellow students were frequently warned against wrecking their health by overusing it for all-night study. In later life, however, he developed an interest in its pharmacology, and Runge impressed him on their first meeting, in which they discussed plant poisons, by dilating a cat's pupil with a drop of belladonna. Goethe gave Runge a handful of Arabian mocha beans to analyse and he succeeded in extracting their pure stimulant essence in his laboratory. Runge dedicated his discovery to the great man and named the new compound caffeine. Other chemists were working along similar lines, and within two years four other German researchers achieved the same result by starting from other caffeine-containing plants such as tea, kola nut and maté.
54. Hanzlik 1938b 141. Nutmeg contains the psychoactive alkaloid myristicin; around 5g of the nut can produce scattered thoughts, distortions of time and space and visual hallucinations. Larger doses produce dizziness, nausea and stupor. The best-known subjective account is in *The Autobiography of Malcolm X* (1965), which describes his use of it as an intoxicant in prison.
55. For Coleridge's nitrous oxide experiences, see Jay 2009 192–4
56. Strickland 1998 462
57. De Quincey 1986 [1821] 72
58. Ibid. 75
59. Ibid. 103
60. Tallis 2012 [2002] 5
61. Tony James 1995 102
62. Phillips 2014 74
63. The term 'scientist' was adopted by committee at the British Association for the Advancement of Science in 1834. By the 1840s it had taken hold in the USA, but was not in general use in Europe until the end of the nineteenth century.
64. *Oxford English Dictionary*
65. Von Bibra 1995 [1855] 152–7
66. Breen 2019 146 ff.
67. Jay 2009 182
68. Oreskes 1996 104
69. It has been noted that, of the medicinal plants from the colonial world studied by nineteenth-century scientists, the class that conspicuously lacks self-experimental description is abortifacients. Schiebinger 2005 153 ff.
70. Johnston 1859 [1855] vol. II 193
71. Moreau 1973 [1845] 1
72. Ibid. 17
73. Tony James 1995 102
74. In Byck (ed.) 1974 62
75. Ibid. 62
76. Ibid. 53
77. Ibid. 164
78. Ibid. 60
79. Ibid. 41
80. Ibid. 60
81. Ibid. 62
82. In Andrews and Solomon (eds) 1975 41
83. Moreau 1973 [1845] 27
84. In Byck (ed.) 1974 60
85. James Crichton Browne 1889, in Milnes 2019 ch. 12 (ebook)
86. Milnes 2019 ch. 12
87. Ibid. 102
88. Pendergrast 2013 22

2 PROSTHETIC GODS

1. Byck (ed.) 1974 32
2. Ibid. 73
3. Davy 1800 556. For more on the reasons why Davy's proposal had to wait fifty years before it became the greatest medical breakthrough of the nineteenth century, see Jay 2009 211–15.
4. Spillane 1999 23
5. Friman 1999 85
6. Byck (ed.) 1974 14
7. Ibid. 39–40
8. Ibid. 156
9. Ibid. 71
10. Ibid. 157–8
11. Ibid. 59
12. Ibid. 62
13. Small 2016 11
14. Ibid. 15
15. Ibid. 197
16. Levinstein 1878 7
17. Ibid. 8
18. Hickman 2004 1276
19. Milligan 2005 541
20. Small 2016 4
21. Levinstein 1878 1
22. Hickman 2004 1279
23. Beddoes 1799 27
24. Levinstein 1878 4; Milnes 2009 ch. 12
25. Byck (ed.) 1974 128
26. Ibid. 144
27. Ibid. 125
28. Ibid. 133
29. Ibid. 145
30. Phillips 2014 88
31. In Byck (ed.) 1974 164
32. Holloway 2016 81
33. Ashley 2005 9
34. Wells 1927 927
35. Ibid. 941
36. Ibid.
37. Ibid. 942. Wells's later novel *Tono-Bungay* (1908) takes its name from a medically worthless but immensely lucrative patent health tonic that brings wealth and status – along with unforeseen complications – to its protagonist.
38. *British Medical Journal* 1887 1229
39. Byck (ed.) 1974 173
40. Ibid. 171
41. See Borch-Jacobsen 2000
42. Ibid. 182
43. Byck (ed.) 1974 183
44. Ibid. 184
45. Ibid. 185
46. Andrews and Solomon (eds) 1975 250
47. Byck (ed.) 1974 186–7

48. Andrews and Solomon (eds) 1975 247
49. Ibid. 250
50. Ibid. 251
51. Ibid. 59
52. Ibid. 115
53. Now Park Avenue South, between Madison and 4th.
54. Markel 2012 108
55. Ibid. 233
56. Ibid. 238–9
57. Ibid. 69
58. Lee 1935 2
59. Ibid. 181
60. Ibid. 26–7
61. Ibid. 29
62. Ibid. 30
63. Ibid. 31
64. Ibid. 37
65. Ibid. 105
66. Ibid. 2
67. Ibid. 103
68. Lewin 1998 [1931] 58
69. Markel 2012 239
70. Stevenson had used coca wine prior to writing *Jekyll and Hyde*, and by 1890 he was using stronger preparations of the drug during his chronic invalidity in Samoa. He wrote to his doctor in England: 'I find I can (almost immediately) fight off a cold with liquid extract of coca', and suggested that 'perhaps a stronger exhibition – injections of cocaine for instance' might do the same for influenza (Rankin 2001 [1987] 217). Stevenson's medications also included ergotine, an alkaloid from the toxic and hallucinogenic ergot fungus from which LSD would be synthesised by Albert Hofmann. Ergotine's vasoconstricting properties made it an effective treatment for haemorrhage, the application for which Hofmann was searching.
71. Stevenson 1979 [1886] 83
72. Ibid. 84
73. Ibid. 81
74. Ibid. 9
75. Ibid. 81
76. Ibid. 82
77. Hammack 2005 90
78. Griffiths 1899 199–200
79. Ibid. 208–9
80. Machen 1995 [1895] 111–13
81. Doyle 1928 4
82. Ibid.
83. Doyle 1929 143
84. Ibid. 144
85. Doyle 1928 12
86. Ibid. 809
87. Conti 2015 118
88. Borch-Jacobsen 2000 4–5
89. Byck (ed.) 1974 255
90. Comers 1923 vi–vii
91. Freud was one of several well-known figures to be 'Steinached' in the 1920s and 1930s; another was William Butler Yeats, whose experiments with hashish and peyote in the occult and literary circles of the 1890s are described in chapter 6. Yeats used hashish in

magical practices that conjured Tantric archetypes, and he understood Steinach's procedure in Tantric terms as a 'self-begetting', through which he might generate superabundant energies by reintegrating the male and female aspects of his body. He was highly satisfied, and convinced that the operation restored his potency, both sexual and poetic. See Kimberley Myers (2009), who reads Yeats's late plays *The King of the Great Clock Tower* and *Full Moon in March* (1934–5), in which the protagonist is beheaded, as symbolic dramas of the Steinach procedure.

92. Doyle 1928 1265
93. Freud 1979 [1929] 13
94. Ibid. 15
95. Ibid. 28–9

3 A WORLD OF PURE EXPERIENCE

1. James 1882, in Green (ed.) 2016 93
2. Ibid. 94
3. Ibid. 95
4. Ibid. 94
5. Ibid. 93
6. James 1874 627
7. Richardson 2007 159
8. James 1874 627
9. Partridge 2018 71
10. Blood 1920 vii
11. Ibid. viii
12. Blood 1874 33
13. In ibid., from Alfred Tennyson, 'The Two Voices', written 1833–4, published in *Poems*, 1842
14. Blood 1874 6
15. Ibid. 36–7
16. Davy 1800 488
17. Jay 2009 61–2
18. James 2000 [1892–1909] 314
19. Davy 1800 489
20. Jay 2009 187
21. Ibid. 186
22. *A Dissertation on the Chymical Properties and Exhilarating Effects of Nitrous Oxide Gas and its Application to Pneumatick Medicine*, in Green (ed.) 2016 75
23. Davy 1800 556
24. Smith 1982 35
25. Jay 2011 32
26. In Green (ed.) 2016 84
27. Ibid.
28. Smith 1982 37
29. Grayson 1996 115
30. Smith 1982 54
31. Jay 2011 34–5
32. Ball 2006 187
33. Ibid. 112
34. Smith 1982 58
35. Gabriel 2010 79
36. Ibid.

37. Snow 2008 30
38. Stratmann 2003 38
39. Gordon 1897 108
40. Ole Secher, letter to the *British Medical Journal*, 25 December 1971, 814
41. Gabriel 2010 68
42. Syndicated in the *Monmouthshire Merlin*, 17 July 1847, 4
43. Warren 1849 5
44. Ibid. 35
45. Ibid. 36
46. Ibid. 35
47. Gabriel 2010 85
48. Holmes, letter to Morton, 21 November 1846, *Oxford English Dictionary*
49. Holmes 1879 46–7
50. Russell 1945 123
51. Boon 2002 100
52. Ibid. 98
53. Original in Emily Dickinson, Poems: Packet XX, Fascicle 10 (held as 109b, Houghton Library, Harvard University); first published in *Unpublished Poems* 1935.
54. Blood 1920 216
55. Ibid. viii
56. James 1874 628
57. In Green (ed.) 2016 93
58. Ibid. 97
59. James 2000 [1892–1909] 188
60. Reed 1997 209
61. James 2000 [1892–1909] 77
62. Ibid. 175
63. James 1978 [1910] 1295
64. James 1995 70
65. Maury 1845 14
66. Maudsley, in McCorristine (ed.) 2012 vol. 1 174–5
67. Ibid. 183
68. Ibid. 173
69. Ibid. 185
70. Ibid. 187
71. Blum 2007 72

4 THE UNSEEN REGION

1. McCorristine (ed.) 2012 vol. 1 206
2. Oppenheim 1985 257
3. Richardson 2007 274
4. McCorristine (ed.) 2012 vol. 5 255
5. Myers 1892 444
6. Blum 2007 105
7. Blood 1920 224
8. *Journal of the Society for Psychical Research*, June 1893, 94–5.
9. Ramsay 1893 237
10. Ibid. 239
11. Ibid. 243–4
12. Blood 1920 223–4. William Ramsay was knighted in 1902, after this lecture but before Blood published his response.

13. The BNAS was reincorporated in 1884 as the London Spiritualist Alliance, which is now known as the College of Psychic Studies. Its headquarters are in Queensberry Place, South Kensington, in the building purchased for it by Sir Arthur Conan Doyle, who became its president in 1925.
14. Wyld 1880 120
15. Ibid.
16. Ibid. 127
17. Wyld 1903 71–2
18. Wyld 1880 127
19. Ibid. 123
20. Ibid. 125
21. Ibid. 66
22. Ibid. 132
23. Besant 1897 55
24. Synge 1912 [1897] 207
25. Ibid. 210–12
26. Ibid. 215
27. Hewitt 2015 61
28. Synge 1912 [1897] 215
29. Binet 1890 14
30. Ibid. 9
31. Sommer 2020 8
32. Ellenberger 1994 [1970] 166; Richardson 2007 344
33. Wells 1927 407
34. Ibid. 416–17
35. Richardson 2007 237
36. Ibid. 106
37. The episode and subsequent disputes over the cause of Gurney's death are discussed in Hamilton 2009 166 ff.
38. Jay 2011 38
39. Richardson 1893 456
40. Ibid. 449–50
41. Ibid. 458
42. Maupassant 2000
43. Kurk 1986 286
44. Dunbar 1905 77
45. Maupassant 2000
46. Ibid
47. Ibid.
48. Ibid.
49. Barnes 2019 73
50. Lorrain 2002 [1895] 109–10
51. Lorrain was an inveterate duellist in a society where honour and insult were taken with high seriousness, and a duel was a more dramatic (as well a quicker and cheaper) recourse than suing for libel. Honour could be satisfied, as it was in Lorrain's famous duel with the young Marcel Proust, by both protagonists missing by a mile.
52. Coleridge, *The Friend*, vol. 1, 1809, Essay III
53. Lorrain 2002 [1895] 134–5
54. Ibid. 34
55. James 1985 [1902] 387
56. In Green (ed.) 2016 93
57. James 1985 [1902] 387–8

58. Ibid. 380–1
59. Ibid. 423
60. William Crookes also suffered attacks on his reputation after major scientific figures such as Michael Faraday and John Tyndall condemned séance phenomena as trickery and flim-flam. However, his ground-breaking work on cathode ray tubes, the radiometer and X-rays secured his professional status and he continued to publicly affirm his belief in supernatural forces and a spirit world.
61. James 1911 174
62. James 1890 251
63. James 1985 [1902] 516
64. Partridge 2018 82
65. Everdell 1997 306
66. James 1890 620
67. Kern 2003 [1983] 151
68. Ibid. 19
69. Menand 2001 384
70. Ibid. 394
71. Du Bois 2007 [1903] 8
72. Ibid. 65
73. Ibid. 76
74. James 2000 [1892–1909] 315
75. James 1978 [1910] 173
76. Ibid.
77. Ibid. 189

5 TALES OF THE HASHISH EATERS

1. *Chemist and Druggist*, 16 April 1892, 568
2. Sheppard 2022 169
3. Ibid.
4. *Chemist and Druggist*, 28 January 1893, 106
5. See Irwin 1994 16–19
6. Stevenson 1906 88
7. Preparations of datura have a long history of criminal use as an aid to robbery across the Middle East, India and South America.
8. Guba 2020b 84
9. In his *Gargantua and Pantagruel* (1532–64), François Rabelais gives a description of the mind-altering effects of the mythical 'herb Pantagruelion', which some commentators have identified as cannabis. It seems, though, that this is an ironic rather than a literal description: according to the Rabelais scholar Mikhail Bakhtin, a satire on the extravagant claims of quack medicine sellers, and perhaps a learned reference to Herodotus' story of the Scythians inhaling cannabis smoke by burning the plant on a brazier. Rabelais was familiar with hemp – his father cultivated it on a large scale at his farm in Chinon – and his paean seems to be a comic skit on the absurdity of making such high-flown claims for a mundane plant, rather than evidence for the psychoactive use of cannabis in early modern Europe.
10. Daftary 1994 169
11. Ibid. 164
12. Guba 2020b 42
13. Ibid. 80
14. Moreau 1973 [1845] 1
15. Guba 2020 125

16. O'Shaughnessy 1839 850
17. Ibid. 839
18. Ibid. 840. 'Ten grains' is around 0.7 grams.
19. Ibid. 842
20. Ibid. 844–5
21. Ibid. 849
22. The doses of hashish that Moreau used are hard to estimate. As with all cannabis products during the era, strength and concentration were variable and impossible to standardise. Edible preparations such as *dawamesc* and *majoun* also varied in strength, though the bitterness that Moreau mentions suggests that Aubert-Roche's formula was a potent one. A spoonful of the 'green jam' may have contained 1–3 grams of hashish, a very large dose by today's standards but commensurate with the intense effects described by Moreau, Gautier and others.
23. Moreau 1973 [1845] 7
24. Ibid. 8–10
25. Ibid. 18
26. Guba 2020b 164
27. Moreau 1973 [1845] 27
28. Ibid. 93
29. Ibid. 28
30. James 1884 189
31. Moreau 1973 [1845] 211
32. Ibid.
33. Ibid. 17
34. Ibid. 17
35. James 1995 107
36. Published in the introduction to Charles Baudelaire, *Les Paradis artificiels*, Gallimard 1961 p. 12.
37. Mickel 1969 93
38. Gautier 1975 85
39. Baudelaire and Gautier 2019 62
40. Ibid. 88
41. Ibid. 90
42. Ibid. 93
43. Hammer-Purgstall 1835 138
44. Davenport-Hines 2001 61
45. Gautier 1915 70
46. Letter to Madame Hanska, 23 December 1845 in Baudelaire 1996 [1860] xv
47. Gautier 1915 71
48. Mickel 1969 83 (my translation)
49. Davenport-Hines 2001 65
50. Dumas 2003 318–20
51. Holmes 1995 [1985] 249
52. Ibid. 247
53. Nerval 1975 107
54. Moreau 1973 [1845] 19
55. James 1995 131
56. Gautier 1915 74–5
57. Guba 2020b 166
58. Samorini 1996 37
59. Gustave Flaubert was among Gastinel's customers, though he may have been too afraid to try the sample that he bought from him (see Schivelbusch 1993 204).

60. Roger wrote under the name Roger de Beauvoir.
61. Gautier 1915 70
62. Baudelaire 1996 [1860] viii
63. De Quincey 1986 [1821] 33
64. Baudelaire 1996 [1860] 38–9
65. Ibid. 40
66. Ibid. 42
67. Ibid.
68. Ibid. 50
69. Ibid. 57
70. Ibid. 71
71. Ibid. 74
72. Ibid. 71
73. Ibid. viii
74. See Hilton 2004
75. Gautier 1915 77
76. Ibid. ix
77. Baudelaire 2010 [1863] 13
78. Calinescu 1987 47
79. Barnes 2019 43

6 EXTASIA, FANTASIA AND ILLUMINATI

1. Deveney 1997 21
2. Ibid. 8
3. Cahagnet 1855 69
4. Cahagnet 1851 116–17
5. Ibid.
6. Ibid. 120
7. Ibid. 122
8. Ibid. 184
9. Ibid. 126
10. Hahnegraaff 2016 108
11. James 1995 107
12. Deveney 1997 146
13. Ibid. 147
14. Taylor 1855 137
15. Ibid. 138
16. Ibid. 141
17. Ludlow 1857 43
18. Johnston 1859 [1855] vol. II 113
19. Dulchinos 1998 38
20. Ibid. 72
21. Ibid. 78
22. Ludlow 1857 91
23. Deveney 1997 70
24. Godwin 1994 260–1
25. Randolph 1867 4
26. Ibid. 1
27. Ibid. 2–4
28. Deveney 1997 72

29. Randolph's biographer John Deveney makes a heroic attempt (Deveney 1997, Appendix C 343–68). The task might be compared to that of assembling a full discography for the musician Sun Ra, on whom Randolph was a spiritual influence.
30. Godwin 1994 258
31. Randolph 1867 2–4
32. Hahnegraaff 2016 113
33. Lévi 2017 136
34. Partridge 2018 376 n3.
35. Until the arrival of adrenaline in the early twentieth century, these toxic fumigants were the most common and effective remedies for asthma or shortness of breath, and from the 1850s were sold as branded patent cigarettes or powders for burning at home in small braziers. Medicinal cigarettes of this kind might also contain opium as a cough suppressant; the product range of the leading French manufacturer, Grimault & Co., included cannabis, mercury and arsenic. The instant relief these products offered to asthma sufferers made them a staple of every pharmacy, and the acrid smoke of deliriant nightshades – along with the distinctive odours of camphor, cannabis and chloroform – a familiar medical presence in the home. Narcotic fumigants feature explicitly in the era's occult fiction, including Edward Bulwer Lytton's *A Strange Story* (1861) and Emma Hardinge Britten's *Ghost Land* (1876). Both these novels have been claimed as *romans à clef* of a real-life society known to their authors as the Orphic circle, in which hallucinogenic smoke was allegedly used in conjunction with crystal gazing for clairvoyance.
36. Blavatsky 2012 [1877] vol. 2 589
37. Deveney 1997 254
38. *Light, A Journal of Psychical, Occult and Mystical Research*, 20 May 1893 233
39. A possible exception was the Hermetic Brotherhood of Luxor, a small British society founded in 1884 and inspired directly, though posthumously, by Paschal Beverly Randolph. For reports that intoxicating potions were used as part of the initiation into its order, see Godwin et al. 2000.
40. Yeats 1922 73
41. Partridge 2018 170
42. Alford 1980 v
43. Ibid. 63
44. Munro 1969 74
45. Symons 1918
46. Alford 1980 58
47. Ibid.
48. Yeats 1907 25–7. The Martinist Order was founded by the Illuminist mystic Louis Claude de St Martin (1743–74). His writings were spread across revolutionary France by 'Societies of the Unknown Philosopher' that formed to study his teachings. These evolved into a Martinist Order, whose president in the 1890s was the Parisian physician and Rosicrucian occultist Papus (Gérard Encausse). In his *Traité élémentaire de magie pratique* (1893), Papus cautioned against the use of hashish in magical practice: 'This substance . . . acts on the energy reserves of the nervous centres, emptying them in an instant.'
49. Ibid. Max Nordau's *Degeneration* (1892) was an influential work of cultural criticism that diagnosed decadence as a pathological symptom of modern life.
50. Yeats 1922 13
51. Coote 1997 153
52. Jeffares 1949 109
53. Gonne 1995 [1937] 257
54. Greer 1995 188
55. Ibid. 187; Berridge 1988 54
56. Yeats's handmade *tattva* cards are held at the National Library of Ireland.

57. Gonne 1995 [1937] 250
58. Ibid. 251
59. Ibid. 259
60. Schmitt 1994 171
61. Symons 2014 [1899] 5
62. Ibid. 8
63. Other plants that would later be included in the grouping 'psychedelic' were being discovered by scientists and explorers at this time. In 1851 the British botanist Richard Spruce encountered ayahuasca during his Amazon travels; he ingested some, but experienced only severe nausea. In 1858 the Ecuadorian geographer Manuel Villavicencio wrote of his visionary experiences after drinking the brew with the Shuar people on the Rio Napo; he reported a heightening of the senses and the sensation of flying to marvellous places (Villavicencio 1858 374). In 1896 the German explorer Theodor Koch-Grünberg took ayahuasca in the Rio Negro region and described hashish-like hallucinations. It was not until the 1920s, however, that the botany and pharmacology of ayahuasca were studied by Western investigators.
64. Ellis 1898 7–8
65. See Jay 2019 6–7
66. Ellis 1898 131. In 1936 the poet and dramatist Antonin Artaud visited the Tarahumara country of northern Mexico to take peyote. His account of the experience was published as *The Peyote Dance*.
67. Ellis 1898 134–5
68. Ibid.
69. Ibid. 139
70. Brewster 1858 [1819] 135
71. Baudelaire 2010 [1863] 13
72. Ellis 1898 138
73. Symons 1918
74. Baudelaire and Gautier 2019 xvii
75. Jewanski et al. 2020 182 ff.
76. Dann 1998 33
77. Ibid. 187
78. Gonne and Yeats 1992 70
79. See Jay 2019 98–100
80. Yeats 1907 27
81. Shiel 1928 [1895] 4–5
82. Ibid.
83. Barnes 2019 37
84. Schmitz 2018 [1902] ix
85. Ibid. 97
86. Ibid. 15
87. Coakley 2017 [1897] 70
88. Kane 1883 946–7
89. Globe-shaped cups or gourds and silver straws are traditionally used in Argentina and Uruguay for drinking maté tea. Coca is traditionally chewed in the Andes, rather than drunk as a tea: as Freud established, cocaine is partially broken down in the stomach, and is more efficiently absorbed into the bloodstream via the mucous membranes.
90. Lee 1935 2
91. Ibid. 60
92. Ibid. 65
93. Eberhardt 1988 [1920] 72
94. Ibid. 74

7 A SIN, A CRIME, A VICE OR A DISEASE?

1. *Chemist and Druggist*, 23 December 1899, 1010–11
2. Parascandola 1997 162
3. Mill 2011 [1859] 141
4. Ibid. 146
5. Putnam and Garrett 2020 168
6. Kerr 1884 1
7. Rasmussen 2008 8
8. Spillane 1999 29
9. Musto 1999 23
10. Wiley had already advocated for a federal ban on the use of the peyote cactus by Native Americans, maintaining that it was a deliriant poison (see Jay 2019).
11. Müller et al. 2006 135
12. Kraepelin 1970 70–1
13. Breathnach 1993 413
14. Rivers 1908 19–20
15. Ibid. 109
16. Ibid. 114
17. Fancher 1996 298
18. Ibid. 301
19. Asprem 2018 177
20. *New York Times*, 25 October 1920, 8
21. Gautier 1915 69
22. Guba 2020b 184
23. Baker 2001 58
24. Schmied et al. 2006 147
25. Parssinen 1983 93
26. Carstairs 2000 54
27. Du Bois 2007 [1903] 3; see Koram (ed.) 2019 for the continuation of the colour line into global drug policy today, and Michelle Alexander's *The New Jim Crow* (2011) for its perpetuation in contemporary US policies of mass incarceration.
28. *New York Times*, 20 March 1905, 14
29. Racial panic was one of several reasons for cocaine's disappearance. The health-conscious public had begun to avoid it in the 1890s; after 1899 the introduction of novocaine, a non-intoxicating local anaesthetic, marginalised its use in dentistry and surgery. In 1903 the American Pharmaceutical Association closed ranks, mandating its removal from patent medicines (including Coca-Cola, which switched to a decocainised coca leaf extract known commercially as 'Merchandise No. 5').
30. Judd 1910 77
31. Kern 2003 [1983] 182
32. Luhan 1936 277
33. Stewart 1987 237
34. Huxley 1932 229
35. Huxley 1980 13
36. Huxley 1932 84
37. In 'Wanted: A New Pleasure', an essay written in 1931 to accompany *Brave New World*, Huxley considers 'a more efficient and less harmful substitute for alcohol and cocaine', but concludes that the 'one genuinely modern pleasure' is the 'inebriating' effect of travelling at high speed (Huxley 1980 9).
38. Regardie 1994 [1968] 100, 106
39. Ibid. 103
40. Ibid.

41. Ibid. 102
42. Ibid. 114–15
43. Ibid. 98. Crowley persisted with the obscure term 'anhalonium' for peyote long after it had been discarded by botanists and pharmacologists. For more on his use of peyote, see Jay 2019 105–9.
44. Partridge 2018 168
45. Ibid. 170
46. Gootenberg 2008 200
47. Davenport-Hines 2001 167
48. Kamieński 2016 100
49. Ibid. 97
50. Lee 1935 212
51. Ibid. 213
52. Ibid. 214
53. Ibid. 215
54. Ibid. 217
55. Coroner's Report, Mulki Reuti Gwalin, City of London, 30 November 1914, No. 154
56. *The Times*, 26 November 1914
57. Lee 1935 25
58. Ibid. 3. On retirement James Lee moved back to the north-east of England, where he lived with his sister in Redcar. He died in 1951, aged 77; his death certificate gave the cause as carcinoma of the liver. His only subsequent book was a short self-published pamphlet entitled *Human Actions Produce Equal Re-Actions: The Secret of Perfect Health* (1946), a self-help tract that drew on the principles of *karma* he had first encountered while staying in a Buddhist temple in China in 1910.

8 TWICE-BORN

1. UN Single Convention 1961, Article 36, 'Penal Provisions'.
2. *The American Magazine*, July 1937; reprinted in *Reader's Digest* 1938.
3. Ibid. The article served as the basis for a 1937 propaganda film of the same name.
4. Guba 2020a 3–4
5. See Lines 2010 8–9. Rick Lines has subsequently discovered that 'evil' is used in one other UN document, the 1948 UN Convention for the Suppression of the Traffic in Persons & of the Exploitation of the Prostitution of Others.
6. The Roman Catholic Church advocated a strict prohibition on all drugs: the use of anaesthetics in surgery had only been formally permitted by Pope Pius XII in 1957 (*Address to an International Congress of Anesthesiologists*, Official Documents, Pope Pius XII, *L'Osservatore Romano*, 24 November 1957).
7. Larkin, 'Annus Mirabilis', 1967, published in *High Windows* (1974).
8. Mills 2013 117
9. Bowles 1962 9–11
10. Ibid. 9
11. Torgoff 2004 61; see also Jay 2019 227–8
12. Hecht 2020 [1954] 15
13. Roszak 1968 xxiii–xxiv
14. Burroughs 1977 [1953] 7
15. New York Public Library, William S. Burroughs archive, Folio 36, Item 4, 'Forward [*sic*] to James Lee's Book', 3rd draft, 4 pp., 9 June 1972
16. Burroughs 1977 [1953] 145–6
17. Burroughs and Ginsberg 2008 [1963] 70
18. Altman 1986 225

19. Haldane 1940 [1927] 106
20. Ibid. 111
21. Ibid. 113
22. Steinberg never took LSD, which she declined to study because she believed it had no medical applications. Obituary, *Guardian*, 21 January 2020.
23. Rasmussen 2008 6
24. Ibid. 18
25. 4.6.2.1.3 Benzedrine Memos and Reports, GSK Heritage Archives
26. Sargant 1967 45
27. Ibid. 46
28. Schultes 1939
29. Graves 1962 135
30. Ibid. 136–7
31. Ibid. 138–9
32. Becker 1953 238
33. I discuss a well-documented example from 1799 in Jay 2011 160–2.
34. Letcher 2006 104
35. Graves 1960 40
36. Graves 1958 327
37. Ibid. 343
38. Albert Hofmann Archives, N. Hofmann 148.1, 22.4.43, Institut für Medizingeschichte der Universität Bern (my translation). I discuss Hofmann's subsequent experiments and other possible sources of the account in his 1979 memoir in https://mikejay.net/bicycle-day-revisited/
39. Krieg 1964 357
40. Hofmann 2013 21
41. Huxley 1980 29
42. Ibid. 16
43. Ibid. 24
44. Ibid. 25
45. Huxley 2004 [1954] 50
46. Huxley 1980 107
47. Ellis 1898 13
48. Huxley 2004 [1954] 14
49. Huxley 1980 36
50. Maslow 1943 393
51. Richards 2016 321
52. https://www.vulture.com/2017/06/cary-grants-lsd-therapy-the-inside-story.html
53. James 1985 [1902] 166
54. The experiments in which Kesey participated are often described as 'CIA funded'. The CIA's MK-Ultra project funded many trials of LSD at this time, including several that violated the norms of consent and human rights. Often the funding was provided through philanthropic and scientific foundations that concealed the CIA's involvement. In this case, Leo Hollister was introduced to psychedelic research at a 1959 conference organised by the Josiah Macy Foundation, who were at that time closely linked to the CIA, but the declared funding for his project came from the National Institute of Mental Health in Bethesda, Maryland.
55. Hollister 1962 80
56. Ibid. 88
57. Huxley 2004 [1954] 3
58. Graves 1962 135
59. Leary 1983 95
60. Lee and Shlain 1985 82
61. Stevens 1989 228
62. Higgs 2006 52

63. Lattin 2010 97
64. Shulgin and Shulgin 1991 16
65. MDMA had been synthesised in 1912 by Merck Pharmaceuticals, while they were working on the synthesis of mescaline, but was never brought to market.
66. Shulgin and Shulgin 1991 xxi
67. Shulgin et al. 1986 315–16
68. The Shulgin Rating Scale was adopted by underground chemists and became a *lingua franca* on the bulletin boards and newsgroups of the early internet era. It thrives today across the global network of informal experimenters with 'research chemicals'. Authorised drug science has subsequently developed its own protocols and questionnaires, such as the OAV Altered States of Consciousness Ratings Scale which asks subjects to rate their experience along dimensions that include 'Oceanic Boundlessness' and 'Visionary Restructuralisation'; but all such quantitative frameworks import the assumptions of their designers (see Baggott 2018 479). Shulgin's system is simple, resilient and culturally universal.

EPILOGUE AFTER DRUGS

1. Verified Market Research, Report ID 37661, 'Nootropics Market Size and Forecast', August 2021, https://www.verifiedmarketresearch.com/product/nootropics-market/
2. It is notable that none of the twenty-first-century revivals of Sherlock Holmes have retained his cocaine-injecting habit: modern film and TV adaptations of nineteenth-century sources typically look for ways to make them more contemporary and 'edgy', but Holmes's habit still needs to be toned down. In some adaptations it is replaced with references to a spell in detox, in others with a nicotine-patch dependency. The only cocaine-injecting protagonist in recent years has been in the HBO series *The Knick* (two seasons, 2014–15), which is set in 1900. The leading character is a cocaine-addicted surgeon whose character draws on William Halsted, and Halsted himself appears in one episode.
3. See, for example, Gallimore 2019 and St John 2015. A notable twentieth-century example of the use of nitrous oxide to generate mystical experiences was the guru Bhagwan Shree Rajneesh, who installed a dental anaesthetic dispenser in his private quarters at Rajneeshpuram, his ashram in Oregon in the 1980s. He dictated a series of books while under the influence of the gas, including *Notes of a Madman* (Osho Books 1985). 'Under the influence of laughing gas', Rajneesh said, 'I am so relieved that I don't have to pretend to be enlightened anymore.' The episode is discussed, with recordings of Rajneesh, in the podcast *Building Utopia*, https://www.buildingutopiapodcast.com/minisode-3
4. The 'major psychedelics' are usually considered to be the tryptamines (LSD, psilocybin, DMT) and phenethylamines (mescaline) that act powerfully on the 5-HT receptors and the serotonin system. However, other substances such as salvinorin, ketamine and phencyclidine (PCP), which act on different neurotransmitter systems, are routinely included in the category, and MDMA, despite being a phenethylamine and powerful serotonin agonist, is often excluded.
5. Hart 2021 190
6. Pollan 2021 86 quotes a total of 1,247,713 for 1997 and 1,239,909 for 2019.
7. Hart 2021 178
8. Ibid. 205
9. Ibid. 246
10. Forstmann and Sagioglou 2021
11. Langlitz 2013 108–9
12. Preciado 2013 [2008] 55
13. Ibid. 250–1
14. Ibid. 353
15. Ibid. 357

BIBLIOGRAPHY

PRINTED BOOKS, ARTICLES AND PAPERS

Archive and manuscript sources, official documents and newspaper references are given in the endnotes

Adams, Clive E., 'James Crichton Browne and Controlled Evaluation of Drug Treatment for Mental Illness', *Journal of the Royal Society of Medicine* 103(4), April 2010, 160–1.

Aday, Jacob S. et al., 'Beyond LSD: A Broader Psychedelic Zeitgeist during the Early to Mid-20th Century', *Journal of Psychoactive Drugs* 51(3), 2018, 210–17.

Alford, Norman, *The Rhymers' Club: Poets of the Tragic Generation*, Cormorant Press, 1980.

Altman, Lawrence K., *Who Goes First? The Story of Self-Experimentation in Medicine*, University of California Press, 1986.

Álvaro, Luis-Carlos, 'Hallucinations and Pathological Visual Perceptions in Maupassant's Fantastical Short Stories – A Neurological Approach', *Journal of the History of the Neurosciences* 14(2), 2006, 100–15.

Aminoff, Michael J., 'The Life and Legacy of Brown-Séquard', *Brain* 140, 2017, 1525–32.

Andrews, George, and David Solomon (eds), *The Coca Leaf and Cocaine Papers*, Harcourt Brace Jovanovich, 1975.

Arnold, Matthew, *Culture and Anarchy*, Smith, Elder & Co., 1869.

Aschenbrandt, T., 'Die physiologische Wirkung und Bedeutung des Cocain. muriat. auf den menschlichen Organismus. Klinische Beobachtungen während der Herbstwaffenübungen des Jahres 1883', *Deutsche Medicinische Wochenschrift* 9(50), 1883, 730–2.

Ashley, Mike, *The Age of the Storytellers*, Oak Knoll, 2005.

—— *Adventures in the Strand*, British Library, 2016.

Asprem, Egil, *The Problem of Disenchantment*, State University of New York Press, 2018.

Baggott, M.J., 'Seed Crystal: The Contributions of Alexander Shulgin to the Science of Consciousness', in A. Shulgin and A. Shulgin, *The Commemorative Edition of Pihkal and Tihkal*, ed. J. Marker, Transform Press, 2018, 474–82.

Baker, Phil, *The Dedalus Book of Absinthe*, Dedalus Press, 2001.

—— *City of the Beast: The London of Aleister Crowley*, Strange Attractor Press, 2022.

Ball, Philip, *The Devil's Doctor: Paracelsus and the Renaissance World of Magic and Science*, Heinemann, 2006.

Balzac, Honoré de, *Treatise on Modern Stimulants*, Wakefield Press, 2018 [1839].

Barnard, William G., *Exploring Unseen Worlds: William James and the Philosophy of Mysticism*, State University of New York Press, 1997.

Barnes, Julian, *The Man in the Red Coat*, Jonathan Cape, 2019.

Barton, William, *A Dissertation on the Chymical Properties and Exhilarating Effects of Nitrous Oxide Gas and its Application to Pneumatick Medicine*, doctoral thesis, University of Pennsylvania, 1808.

Barzun, Jacques, *A Stroll with William James*, University of Chicago Press, 1983.

Baudelaire, Charles, *Artificial Paradises*, trans. Stacy Diamond, Citadel Press, 1996 [1860].

—— *The Painter of Modern Life*, trans. P.E. Charvet, Penguin, 2010 [1863].

Baudelaire, Charles, and Théophile Gautier, *Hashish, Wine, Opium*, trans. Maurice Stang, Alma Classics, 2019.

Beard, George M., *A Practical Treatise on Nervous Exhaustion (Neurasthenia)*, William Wood & Co., 1880.

—— *American Nervousness: Its Causes and Consequences*, G.P. Putnam's Sons, 1881.

Becker, Howard S., 'Becoming a Marihuana User', *American Journal of Sociology* 59(3), November 1953, 235–42.

—— *Outsiders: Studies in the Psychology of Deviance*, The Free Press, 1963.

Beddoes, Thomas, *Notes of Some Observations Made at the Medical Pneumatic Institution*, Biggs & Cottle, 1799.

Benjamin, Walter, 'On Some Motifs in Baudelaire', *Selected Writings*, vol. 4: *1938–40*, Harvard University Press, 2003.

—— *On Hashish*, trans. Howard Eiland, Belknap Press, 2006.

Bentley, W.H., 'Erythoxylon Coca in the Opium and Alcohol Habits', *Detroit Therapeutic Gazette*, 15 September 1880.

Bergson, Henri, *Time and Free Will*, Dover Publications, 2001 [1889].

Beringer, Kurt, *Der Meskalinrausch*, Springer-Verlag, 1969 [1927].

Berridge, Kent C., and Morten L. Kringelbach, 'Building a Neuroscience of Pleasure and Well-Being', *Psychology of Well-Being: Theory, Research and Practice* 1, 2011, Article 3.

Berridge, Virginia, 'The Origins of the English Drug "Scene"', *Medical History* 32, 1988, 51–64.

Berrios, German, *History of Mental Symptoms*, Cambridge University Press, 1996.

Besant, Annie, *The Ancient Wisdom: An Outline of Theosophical Teachings*, Theosophical Publishing House, 1897.

Bewley-Taylor, David, and Martin Jelsma, 'Regime Change: Revisiting the 1961 Single Convention on Narcotic Drugs', *International Journal of Drug Policy* 23(1), January 2012, 72–81.

Biden, Hunter, *Beautiful Things*, Gallery Books, 2021.

Binet, Alfred, *On Double Consciousness: Experimental Psychological Studies*, Open Court Publishing Co., 1890.

Bjelić, Dušan, *Intoxication, Modernity and Consumerism*, Palgrave Macmillan, 2017.

Blavatsky, Helena P., *Isis Unveiled: A Master-Key to the Mysteries of Ancient and Modern Science and Theology*, Cambridge University Press, 2012 [1877].

Blood, Benjamin Paul, *The Anaesthetic Revelation and the Gist of Philosophy*, Amsterdam, New York 1874.

—— *The Pluriverse*, Marshall Jones Co., 1920.

Blum, Deborah, *Ghost Hunters*, Century, 2007.

Boon, Marcus, *The Road of Excess: A History of Writers on Drugs*, Harvard University Press, 2002.

Borch-Jacobsen, Mikkel, 'How a Fabrication Differs from a Lie', *London Review of Books*, 13 April 2000.

Boren, Lynda S., 'William James, Theodore Dreiser and the Anaesthetic Revelation', *American Studies* 42(1), 1983, 5–17.

Bowles, Paul, *A Hundred Camels in the Courtyard*, City Lights Books, 1962.

Breathnach, C.S., 'W.H.R. Rivers and the Hazards of Interpretation', *Journal of the Royal Society of Medicine*, 86, July 1993, 413–16.

Breen, Benjamin, 'Victorian Occultism and the Art of Synaesthesia', *Public Domain Review*, 19 March 2014.

—— *The Age of Intoxication*, University of Pennsylvania Press, 2019.

Brewster, Sir David, *The Kaleidoscope: Its History, Theory and Construction*, John Murray, 1858 [1819].

Brown-Séquard, C., 'The Effects Produced on Man by Subcutaneous Injections of a Liquid Obtained from the Testicles of Animals', *The Lancet* 2, 1889, 105–7.

Burroughs, William, *Junky*, Penguin Books, 1977 [1953].

—— and Allen Ginsberg, *The Yage Letters*, Penguin Classics, 2008 [1963].

Byck, Robert (ed.), *Cocaine Papers: Sigmund Freud*, Stonehill Publishing, 1974.

Cahagnet, Louis-Alphonse, *The Sanctuary of Spiritualism*, trans. M. Flinders Pearson, Geo. Peirce, 1851.

—— *The Celestial Telegraph*, Partridge & Brittan, 1855.

Calinescu, Matei, *Five Faces of Modernity: Modernism, Avant-Garde, Decadence, Kitsch, Postmodernism*, Duke University Press, 1987.

Canales, Jimena, *The Physicist and the Philosopher*, Princeton University Press, 2015.

Caquet, P.E., *Opium's Orphans: The 200-Year History of the War on Drugs*, Reaktion Books, 2022.

Carstairs, Catherine, 'The Most Dangerous Drug: Images of African-Americans and Cocaine Use in the Progressive Era', *Left History* 7(1), March 2000, 46–61.

Chambers, John Whiteclay, *The Tyranny of Change: America in the Progressive Era*, Rutgers University Press, 1992.

Church, Roy, and E.M. Tansey, *Burroughs Wellcome & Co.*, Crucible Books, 2007.

Coakley, T.W., *Keef: A Story of Intoxication, Love and Death*, ed. Ronald K. Siegel, Process Media, 2017 [1897].

Comers, George F., *Rejuvenation: How Steinach Makes People Young*, Thomas Seltzer, 1923.

Connor, Steven, *The Matter of Air: Science and Art of the Ethereal*, Reaktion Books, 2010.

Conti, Meredith, 'Ungentlemanly Habits: The Dramaturgy of Drug Addiction in Fin-de-Siècle Theatrical Adaptations of the Sherlock Holmes Stories and *The Strange Case of Dr Jekyll and Mr Hyde*', in *Victorian Medicine and Popular Culture*, ed. Louise Penner and Tabitha Sparks, University of Pittsburgh Press, 2015.

Coote, Stephen, *W.B. Yeats: A Life*, Hodder & Stoughton, 1997.

Courtwright, David, *Forces of Habit: Drugs and the Making of the Modern World*, Harvard University Press, 2001.

—— *The Age of Addiction: How Bad Habits Became Big Business*, Belknap Press, 2019.

Cowles, Henry, *The Scientific Method: An Evolution of Thinking from Darwin to Dewey*, Harvard University Press, 2020.

Crary, Jonathan, 'Techniques of the Observer', *October* 45, Summer 1988, 3–35.

—— *Techniques of the Observer: On Vision and Modernity in the Nineteenth Century*, MIT Press, 1991.

—— *Suspensions of Perception: Attention, Spectacle and Modern Culture*, MIT Press, 1999.

Critchley, Macdonald, 'Mescalism', *British Journal of Inebriety* 28, January 1931, 100–8.

Daftary, Farhad, *The Assassin Legends: Myths of the Isma'ilis*, I.B. Tauris & Co., 1994.

Dann, Kevin T., *Bright Colours Falsely Seen: Synesthesia and the Search for Transcendental Knowledge*, Yale University Press, 1998.

Daston, Lorraine, and Peter Galison, *Objectivity*, Zone Books, 2007.

Davenport-Hines, Richard, *The Pursuit of Oblivion*, Weidenfeld & Nicolson, 2001.

Davis, Wade, 'Smoking Toad', *Shadows of the Sun*, Island Press, 1998.

—— and Andrew Weil, 'Identity of a New World Psychoactive Toad', *Ancient Mesoamerica* 3(1), March 1992, 51–9.

Davy, Humphry, *Researches Chemical and Philosophical, Chiefly Concerning Nitrous Oxide and its Respiration*, Joseph Johnson, 1800.

De Quincey, Thomas, *Confessions of an English Opium-Eater*, Penguin Classics, 1986 [1821].

Deveney, John Patrick, *Paschal Beverly Randolph*, State University of New York Press, 1997.

Devenot, Neše, *Altered States/Other Worlds: Romanticism, Nitrous Oxide and the Literary Prehistory of Psychedelia*, dissertation, University of Pennsylvania, 2015.

Donlon, Helen, *Shadows across the Moon: Outlaws, Freaks, Shamans and the Making of Ibiza Clubland*, Jawbone Press, 2017.

Dowbiggin, Ian, 'Alfred Maury and the Politics of the Unconscious in Nineteenth-Century France', *History of Psychiatry* 1, 1990, 255–87.

Doyle, Arthur Conan, *Sherlock Holmes: The Complete Short Stories*, John Murray, 1928.

—— *Sherlock Holmes: The Complete Long Stories*, John Murray, 1929.

Du Bois, W.E.B., *The Souls of Black Folk*, Oxford University Press, 2007 [1903].

Du Plessis, Michael, 'Unspeakable Writing: Jean Lorrain's Monsieur de Phocas', *French Forum* 27(2), Spring 2002, 65–98.

Dulchinos, Donald P., *Pioneer of Inner Space: The Life of Fitz Hugh Ludlow, Hasheesh Eater*, Autonomedia, 1998.

Dumas, Alexandre, *The Count of Monte Cristo*, trans. R. Buss, Penguin, 2003.

Dunbar, Ernest, 'The Light Thrown on Psychological Processes by the Action of Drugs', *Proceedings of the Society for Psychical Research* 16, 1905, 62–77.

Earleywine, Mitch (ed.), *Mind-Altering Drugs: The Science of Subjective Experience*, Oxford University Press, 2005.

Eberhardt, Isabelle, *The Oblivion Seekers*, trans. Paul Bowles, Peter Owen, 1988 [1920].

Ellenberger, Henri F., *The Discovery of the Unconscious*, Fontana Press, 1994 [1970].

Ellis, Havelock, 'A New Artificial Paradise', *Contemporary Review* 73, January 1898, 130–41.

Enright, Damien, *Dope in the Age of Innocence*, Liberties Press, 2010.

Estes, J. Worth, 'John Jones' *Mysteries of Opium Reveal'd* (1701): Key to Historical Opiates', *Journal of the History of Medicine and Allied Sciences* 34(2), April 1979, 200–10.

Evans, Jules, 'Abraham Maslow, Empirical Spirituality and the Crisis of Values', *Medium*, April 2021, https://julesevans.medium.com/abraham-maslow-empirical-spirituality-and-the-crisis-of-values-34148b775f1a.

Everdell, William R., *The First Moderns*, University of Chicago Press, 1997.

Fancher, Raymond, *Pioneers of Psychology*, 3rd edn, W.W. Norton & Co., 1996.

Ferris, Paul, *Dr Freud: A Life*, Sinclair-Stevenson, 1997.

Fiks, Arsen, *Self-Experimenters: Sources for Study*, Praeger, 2003.

Forstmann, Matthias, and Christina Sagioglou, 'How Psychedelic Researchers' Self-Admitted Substance Use and their Association with Psychedelic Culture Affect People's Perceptions of their Scientific Integrity and the Quality of their Research', *Public Understanding of Science* 30(3), 2021, 302–18.

Foster, Hal, *Prosthetic Gods*, MIT Press, 2006.

Franklin, Dr Benjamin et al., *Animal Magnetism: Report of Dr. Franklin and Other Commissioners, Charged by the King of France with the Examination of the Animal Magnetism as Practised at Paris*, H. Perkins, 1837 [1785].

Freud, Sigmund, *Cocaine Papers*, see Byck, Robert (ed.).

—— *The Complete Letters of Sigmund Freud to Wilhelm Fliess 1887–1904*, see Masson, M. (ed.) 1985.

—— *Civilization and its Discontents*, Hogarth Press, 1979 [1929].

Friman, H. Richard, 'Germany and the Transformation of Cocaine 1880–1920', in *Cocaine, Global Histories*, ed. P. Gootenberg, Routledge, 1999, 83–104.

Fuller, Robert C., 'Biographical Origins of Psychological Ideas: Freud's Cocaine Studies', *Journal of Humanistic Psychology* 32(3), Summer 1992, 67–86.

Gabriel, Joseph M., 'Anaesthetics and the Chemical Sublime', *Raritan*, 30(1), Summer 2010, 68–93.

Gallimore, Andrew, *Alien Information Theory: Psychedelic Drug Technologies and the Cosmic Game*, Strange World Press, 2019.

Gault, Alan, *The Founders of Psychical Research*, Schocken, 1968.

Gautier, Théophile, 'Le Haschisch', *Annales Médico-psychologiques* 11, November 1833, 492.

—— *Charles Baudelaire: His Life*, Greening & Co., 1915.

—— 'The Hashish-Eaters' Club' ('Le Club des Haschischins'), in *The Hashish Club*, vol. 1, ed. Peter Haining, Peter Owen Ltd, 1975.

—— *My Fantoms*, ed./trans. Richard Holmes, New York Review of Books, 1976.

—— with Christian Chanel and John P. Deveney, *The Hermetic Brotherhood of Luxor*, Red Wheel/Weiser, 2000.

Godwin, Joscelyn, *The Theosophical Enlightenment*, State University of New York Press, 1994.

Godwin, J., C. Chanel and J.P. Deveney, *The Hermetic Brotherhood of Luxor*, Samuel Weiser, Inc. 1995.

Golinski, Jan, *The Experimental Self: Humphry Davy and the Making of a Man of Science*, University of Chicago Press, 2017.

Gonne, Maud, *A Servant of the Queen: The Autobiography of Maud Gonne*, University of Chicago Press, 1995 [1937].

—— and W.B. Yeats, *The Gonne–Yeats Letters 1893–1938*, ed. Anna MacBride White and A. Norman Jeffares, Hutchinson, 1992.

Gootenberg, P. (ed.), *Cocaine, Global Histories*, Routledge, 1999.

—— *Andean Cocaine: The Making of a Global Drug*, University of North Carolina Press, 2008.

Gordon, H.L., *Sir James Young Simpson and Chloroform*, T. Fisher Unwin, 1897.

Gower, Barry, *Scientific Method*, Routledge, 1997.

Graves, Robert, *Steps: Stories, Talks, Essays, Poems, Studies in History*, Cassell, 1958.

—— *The White Goddess*, 3rd edn, Farrar, Straus & Giroux, 1960.

—— *Oxford Addresses on Poetry*, Doubleday, 1962.

Grayson, Ellen Hickey, 'Social Order and Psychological Disorder: Laughing Gas Demonstrations 1800–1850', in *Freakery: Cultural Spectacles of the Extraordinary Body*, ed. Rosemarie Garland Thomson, New York University Press, 1996.

Green, Adam (ed.), *Oh Excellent Air Bag: Under the Influence of Nitrous Oxide 1799–1920*, Public Domain Review Press, 2016.

Greer, Mary E., *Women of the Golden Dawn*, Park Street Press, 1995.

Griffiths, George, 'A Genius for a Year', *Gambles with Destiny*, F.V. White, 1899.

Guba, David A. Jr (a), ' "A Sovereign Remedy": Grimault & Co.'s Asthma Cigarette Empire', *Points: Blog of the Alcohol and Drugs History Society (ADHS)*, 18 Feb. 2020, https://pointshistory.com/2020/02/18/a-sovereign-remedy-grimault-cos-asthma-cigarette-empire.

—— (b) *Taming Cannabis: Drugs and Empire in Nineteenth-Century France*, McGill-Queens University Press, 2020.

Gurney, Edmund, 'Hallucinations', in *Spiritualism, Mesmerism and the Occult 1800–1920*, vol. 1, ed. Shane McCorristine, Pickering & Chatto, 2012 (first published in *Mind* 10(38), 1885, 171–89).

Hacking, Ian, *Mad Travelers*, Harvard University Press, 2002.

Hahnegraaff, Wouter, 'The First Psychonaut? Louis-Alphonse Cahagnet's Experiments with Narcotics', *International Journal for the Study of New Religion* 7(2), 2016, 105–23.

Haldane, J.B.S., *Possible Worlds*, Evergreen Books, 1940 [1927].

Hall, Wayne, 'Why Was Early Therapeutic Research on Psychedelic Drugs Abandoned?', *Psychological Medicine*, published online, Cambridge University Press, October 2021.

Hamilton, Trevor, *Immortal Longings: F.H.W. Myers and the Victorian Search for Life after Death*, Imprint Academic, 2009.

Hammack, Brenda Mann, 'The Chemically Inspired Intellectual in Occult Fiction', *Mosaic: An Interdisciplinary Critical Journal*, 37(1), March 2004, 83–99.

Hammer-Purgstall, Joseph, *The History of the Assassins*, trans. O.C. Wood, Smith & Elder, 1835.

Hammond, W.A., 'Coca: Its Preparations and their Therapeutic Qualities, with some Remarks on the So-Called "Cocaine Habit"', in *Cocaine Papers: Sigmund Freud*, ed. Robert Byck, Stonehill Publishing, 1974.

Hanzlik, P.J., 'Purkinje's Pioneer Self-Experiments in Pharmacology, Part 1', *California and Western Medicine* 49(1), July 1938a.

—— 'Purkinje's Pioneer Self-Experiments in Pharmacology, Part 2', *California and Western Medicine* 49(2), August 1938b.

Haridas, Rajesh P., 'Horace Wells' Demonstration of Nitrous Oxide in Boston', *Anaesthesiology* 119(5), November 2013, 1014–22.

Harper, George Mills, *Yeats' Golden Dawn*, Macmillan, 1974.

Harrison, John E., *Synaesthesia: The Strangest Thing*, Oxford University Press, 2001.

Hart, Carl, *Drug Use for Grown-Ups*, Penguin Press, 2021.

Hayter, Alethea, *Opium and the Romantic Imagination*, Faber & Faber, 1968.

Healy, David, *The Creation of Psychopharmacology*, Harvard University Press, 2002.

Hecht, Ben, *A Child of the Century*, Yale University Press, 2020 [1954].

Herzberg, David, *White Market Drugs: Big Pharma and the Hidden History of Addiction in America*, University of Chicago Press, 2020.

Herzig, Rebecca, *Suffering for Science: Reason and Sacrifice in Modern America*, Rutgers University Press, 2006.

Hewitt, Seán, '"An Initiated Mystic": Modernisation and Occultism in Synge's "The Aran Islands"', *New Hibernia Review* 19(4), Winter 2015, 58–76.

Hickman, Timothy A., 'Mania Americana: Narcotic Addiction and Modernity in the United States 1870–1920', *Journal of American History* 90(4), March 2004, 1269–94.

Higgs, John, *I Have America Surrounded: The Life of Timothy Leary*, Friday Books, 2006.

Hilton, Frank, *Baudelaire in Chains: A Portrait of the Artist as a Drug Addict*, Peter Owen Ltd, 2004.

Hofmann, Albert, *LSD, My Problem Child*, trans. Jonathan Ott, Oxford University Press, 2013.

Hollister, Leo E., 'Drug-Induced Psychoses and Schizophrenic Reactions: A Critical Comparison', *Annals of the New York Academy of Sciences* 96, January 1962, 80–92.

Holloway, Verity, *The Mighty Healer: Thomas Holloway's Victorian Patent Medicine Empire*, Pen & Sword History, 2016.

Holmes, Oliver Wendell (Sr), *Mechanism in Thought and Morals, a Lecture Delivered before the Phi Beta Kappa Society of Harvard University, June 29 1870*, Houghton, Osgood & Co., 1879.

Holmes, Richard, *Footsteps*, Flamingo, 1995 [1985].

Hooke, Robert, *Philosophical Experiments and Observations*, W. & J. Innys, 1726.
—— *The Diary of Robert Hooke*, ed. H.W. Robinson, Taylor & Francis, 1935.
Hughes-Hallett, Lucy, *The Pike: Gabriele D'Annunzio*, Fourth Estate, 2013.
Hunt, Geoffrey P., and Kristin Evans, 'The Great Unmentionable: Exploring the Pleasures and Benefits of Ecstasy from the Perspectives of Drug Users', *Drugs, Education, Prevention and Policy* 15(4), 2008, 329–49.
Huxley, Aldous, *Brave New World*, Doubleday, 1932.
—— *The Doors of Perception*, Penguin Vintage Classics, 2004 [1954].
—— *Moksha: Writings of Psychedelics and the Visionary Experience 1931–1963*, ed. Michael Horowitz and Cynthia Palmer, Chatto & Windus, 1980.
Huysmans, Joris-Karl, *À Rebours (Against Nature)*, trans. Robert Baldick, Penguin Classics, 2003 [1884].
Irwin, Robert, *The Arabian Nights: A Companion*, Allen Lane, 1994.
Jackson, Mark, ' "Divine Stramonium": The Rise and Fall of Smoking for Asthma', *Medical History* 54(2), 2010, 171–94.
James, Tony, *Dreams, Creativity and Madness in Nineteenth-Century France*, Clarendon Press, 1995.
James, William, 'Review of "The Anaesthetic Revelation and the Gist of Philosophy" ', *Atlantic Monthly* 33(5), November 1874, 627–8.
—— 'On Some Hegelisms', *Mind* 7, 1882.
—— 'What Is an Emotion?', *Mind* 9(34), April 1884, 188–205.
—— *The Principles of Psychology*, Henry Holt & Co., 1890.
—— *The Varieties of Religious Experience*, Penguin, 1985 [1902].
—— *Pragmatism and Other Writings*, Penguin Classics, 2000 [1892–1909].
—— 'A Pluralist Mystic', in *Essays in Philosophy*, ed. Frederick H. Burkhardt, Fredson Bowers and Ignas K. Skrupskelis, Harvard University Press, 1978 [1910].
—— 'Final Impressions of a Psychical Researcher', *Memories and Studies*, Longman, 1911.
Jay, Mike, *The Atmosphere of Heaven*, Yale University Press, 2009.
—— *Emperors of Dreams: Drugs in the Nineteenth Century*, revised edn, Dedalus Press, 2011.
—— 'The Green Jam of "Doctor X": Science and Literature at the Club des Hashischins', in *Literature and Intoxication*, ed. Eugene Brennan and Russell Williams, Palgrave Macmillan, 2015.
—— 'Miracle or Menace? The Arrival of Cocaine 1860–1900', in *The Neuropsychiatric Complications of Stimulant Abuse*, ed. Pille Taba, Andrew Lees and Katrin Sikk, *International Review of Neurobiology* 120, 2015, ch. 3.
—— 'What Is a Drug?', in *The Pharmakon*, ed. Hermann Herlinghaus, Universitätsverlag, Heidelberg, Winter 2018, 21–34.
—— *Mescaline: A Global History of the First Psychedelic*, Yale University Press, 2019.
Jeffares, A. Norman, *W.B. Yeats: Man and Poet*, Routledge & Kegan Paul, 1949.
Jenkins, Philip, *Dream Catchers: How Mainstream America Discovered Native Spirituality*, Oxford University Press, 2004.
Jessop, Walter H., 'Cocaine', *The Practitioner*, January 1885.
Jewanski, Jörg et al., 'The "Golden Age" of Synaesthesia Enquiry in the Late Nineteenth Century (1876–1895)', *Journal of the History of the Neurosciences: Basic and Clinical Perspectives* 29(2), 2020, 175–202.
Johnston, James, *The Chemistry of Common Life*, 2 vols, William Blackwood & Sons, 1859 [1855].
Jones, John, *The Mysteries of Opium Reveal'd*, Richard Smith, 1701.

Judd, C.H., 'Evolution and Consciousness', *Psychological Review* 17(2), March 1910.

Jung, Carl, *Letters*, vol. 2: *1951–1961*, ed. Gerhard Adler, Princeton University Press, 1976.

Kamieński, Łukas, *Shooting Up: A History of Drugs in Warfare*, Hurst & Co., 2016.

Kane, H.H., *Drugs that Enslave: The Opium, Morphine, Chloral and Haschisch Habits*, Presley Blakiston, 1881.

—— 'A Hashish House in New York', *Harper's Monthly*, November 1883.

Kantor, Arlene Finger, 'Upton Sinclair and the Pure Food and Drugs Act of 1906', *American Journal of Public Health* 66(12), 1976, 1202–5.

Karmel, Richard, 'Freud's Cocaine Papers: A Commentary', *Canadian Journal of Psychoanalysis* 11(1), 2003, 161–9.

Kern, Stephen, *The Culture of Time and Space 1880–1918*, Harvard University Press, 2003 [1983].

—— 'Changing Concepts and Experiences of Time and Space', in *The Fin-de-Siècle World*, ed. Michael Saler, Routledge, 2015.

Kerr, Norman, *Inaugural Address: Society for the Study and Cure of Inebriety*, H.K. Lewis 1884.

Klüver, Heinrich, *Mescal and Mechanisms of Hallucination*, University of Chicago Press, 1966 [1930].

Knauer, Alwyn, and William Maloney, 'A Preliminary Note on the Psychic Action of Mescalin', *Journal of Nervous and Mental Disease* 40(7), July 1913, 425–36.

Kohn, Marek, *Narcomania*, Faber & Faber, 1987.

—— *Dope Girls: The Birth of the British Drug Underground*, Lawrence & Wishart, 1992.

Koram, Kojo (ed.), *The War on Drugs and the Global Colour Line*, Pluto Press, 2019.

Kraepelin, Emil, *Memoirs*, Springer-Verlag, 1970.

Krieg, Margaret, *Green Medicine: The Search for Plants that Heal*, Rand McNally, 1964.

Kroker, Kenton, 'The Progress of Introspection in America 1896–1938', *Studies in History and Philosophy of Biomedical Sciences* 34, 2003, 77–108.

Kurk, Katherine C., 'Maupassant and the Divided Self: "Qui Sait?"', *Nineteenth-Century French Studies* 14, Spring/Summer 1986, 284–94.

Langlitz, Nicolas, 'The Persistence of the Subjective in Neuropsychopharmacology', *History of the Human Sciences* 23(1), 2010, 37–57.

—— *Neuropsychedelia*, University of California Press, 2013.

—— et al., 'Moral Psychopharmacology Needs Moral Inquiry: The Case of Psychedelics', *Frontiers in Psychiatry*, 2 August 2021.

Laski, Marghanita, *Ecstasy: A Study of Some Secular and Religious Experiences*, Cresset Press, 1961.

Lattin, Don, *The Harvard Psychedelic Club*, Harper One, 2010.

Leary, Timothy, *Flashbacks: An Autobiography*, Heinemann, 1983.

Lee, James S., *The Underworld of the East*, Sampson, Marston & Low, 1935.

Lee, Martin A., and Bruce Shlain, *Acid Dreams: The CIA, LSD and the Sixties Rebellion*, Grove Press, 1985.

Lefebure, Molly, *Samuel Taylor Coleridge: A Bondage of Opium*, Victor Gollancz, 1974.

Letcher, Andy, *Shroom: A Cultural History of the Magic Mushroom*, Faber & Faber, 2006.

Lévi, Eliphas, *The Doctrine and Ritual of High Magic*, Tarcher Perigee, 2017.

Levinstein, Edward, *Morbid Craving for Morphia*, trans. Charles Harrer, Smith, Elder & Co., 1878.

Lewin, Louis, *Phantastica*, Park Street Press, 1998 [1931].

Lines, Rick, 'Deliver Us from Evil? – The Single Convention on Narcotic Drugs, 50 Years On', *International Journal on Human Rights and Drug Policy* 1, 2010, 3–14.

Lorrain, Jean, *Nightmares of an Ether-Drinker*, trans. Brian Stableford, Tartarus Press, 2002 [1895].

—— *Monsieur de Phocas*, trans. Francis Amery, Dedalus Books, 1994 [1901].

Ludlow, Fitz Hugh, *The Hasheesh Eater: Being Passages from the Life of a Pythagorean*, Harper & Brothers, 1857.

Luhan, Mabel Dodge, *Intimate Memories*, vol. 3: *Movers and Shakers*, Harcourt Brace & Co., 1936.

Luke, David, *Otherworlds: Psychedelics and Exceptional Human Experience*, Aeon Academic, 2019.

Machen, Arthur, *The Three Impostors*, Everyman Library, 1995 [1895].

Maehle, Andreas-Holger, *Drugs on Trial: Experimental Pharmacology and Therapeutic Innovation in the Eighteenth Century*, Editions Rodopi, 1999.

Makari, George, *Revolution in Mind*, Duckworth, 2008.

Mantegazza, Paolo, 'On the Hygienic and Medicinal Values of Coca', in *The Coca Leaf and Cocaine Papers*, ed. George Andrews and David Solomon, Harcourt Brace Jovanovich, 1975 [1859].

Markel, Howard, *An Anatomy of Addiction*, Vintage Books, 2012.

Maslow, Abraham, 'A Theory of Human Motivation', *Psychological Review* 50, 1943, 370–96.

Masson, M. (ed.), *The Complete Letters of Sigmund Freud to Wilhelm Fliess 1887–1904*, Belknap Press, 1985.

Maudsley, Henry, 'Hallucinations of the Senses', in *Spiritualism, Mesmerism and the Occult 1800–1920*, vol. 1, ed. Shane McCorristine, Pickering & Chatto, 2012, 173–89 (first published in *Fortnightly Review* 24, 1878, 370–86).

—— *Natural Causes and Supernatural Seemings*, Cambridge University Press, 2011 [1887].

Maupassant, Guy, *Complete Short Stories*, vol. 11, 2000, https://www.gutenberg.org/files/3090/3090-h/3090-h.htm.

Maury, Alfred, *De l'hallucination au point de vue philosophique et historique*, Paris, 1845.

Mayer-Gross, Wilhelm, 'Experimental Psychoses and Other Mental Abnormalities Produced by Drugs', *British Medical Journal* 2(4727), August 1951, 317–21.

McCorristine, Shane (ed.), *Spiritualism, Mesmerism and the Occult 1800–1920*, 5 vols, Pickering & Chatto, 2012.

Menand, Louis, *The Metaphysical Club*, Harper Collins, 2001.

—— *The Free World*, Fourth Estate, 2021.

Metzinger, Thomas, *Being No-One: The Self-Model Theory of Subjectivity*, MIT Press, 2004.

Micale, Mark (ed.), *The Mind of Modernism: Medicine, Psychology and the Cultural Arts in Europe 1880–1940*, Stanford University Press, 2004.

—— 'France', in *The Fin-de-Siècle World*, ed. Michael Saler, Routledge, 2015.

Mickel, Emmanuel, *The Artificial Paradises in French Literature*, University of North Carolina Press, 1969.

Mill, John Stuart, *On Liberty*, Walter Scott Publishing Co., 2011 [1859].

Milligan, Barry, 'Morphine-Addicted Doctors, the English Opium-Eater, and Embattled Medical Authority', *Victorian Literature and Culture* 33, 2005, 541–53.

Mills, James H., *Cannabis Britannica: Empire, Trade and Prohibition 1800–1928*, Oxford University Press, 2003.

—— *Cannabis Nation*, Oxford University Press, 2013.

Milner, Max, *L'Imaginaire des drogues*, Editions Gallimard, 2000.

Milnes, Christopher, *A History of Euphoria*, Palgrave Macmillan, 2019.

Mogar, Robert E., 'Current Status and Future Trends in Psychedelic (LSD) Research', *Journal of Humanistic Psychology* 5(2), October 1965, 147–66.

Moore, Virginia, *The Unicorn: W.B. Yeats' Search for Reality*, Macmillan, 1954.

Moreau, Jacques-Joseph, *Hashish and Mental Illness*, Raven Press, 1973 [1845].

Most, Albert, *Bufo Alvarius: Psychedelic Toad of the Sonoran Desert*, Venom Press, 1984.

Mrabet, Mohammed, *M'Hashish*, trans. Paul Bowles, Peter Owen Publishers, 1988 [1969].

Mukherjee, Sujaan, *W.B. O'Shaughnessy and the Introduction of Cannabis to Western Medicine*, Public Domain Review, 2017, https://publicdomainreview.org/essay/w-b-o-shaughnessy-and-the-introduction-of-cannabis-to-modern-western-medicine

Müller, Ulrich et al., 'The Origin of Psychopharmacology: Emil Kraepelin's Experiments in Leipzig, Dorpat and Heidelberg (1882–92)', *Psychopharmacology* 184, 2006, 131–8.

Munro, John M., *Arthur Symons*, Twayne Publishers, 1969.

Musto, David F., 'Sherlock Holmes and Sigmund Freud', in *Cocaine Papers: Sigmund Freud*, ed. R. Byck, Stonehill Publishing, 1974.

—— *The American Disease: Origins of Narcotic Control*, 3rd edn, Oxford University Press, 1999.

Myers, Frederick, 'The Subliminal Consciousness', *Proceedings of the Society for Psychical Research* 8, 1892, 436–535.

Myers, Kimberley, 'W.B. Yeats' Steinach Operation, Hinduism, and the Severed-Head Plays of 1934–5', *Literature and Medicine*, 28(1), Spring 2009, 102–37.

Nerval, Gérard de, 'Hashish' (The Tale of the Caliph Hakim), in *The Hashish Club*, vol. 1, ed. Peter Haining, Peter Owen Ltd, 1975.

Nicotra, Jodie, 'William James in the Borderlands: Psychedelic Science and the "Accidental Fences" of Self', *Configurations* 16(2), Spring 2008, 199–213.

Oppenheim, Janet, *The Other World: Spiritualism and Psychical Research in England 1850–1914*, Cambridge University Press, 1985.

—— *Shattered Nerves: Doctors, Patients and Depression in Victorian England*, Oxford University Press, 1991.

Oram, Matthew, 'Efficacy and Enlightenment: LSD Psychotherapy and the Drug Amendments of 1962', *Journal of the History of Medicine and Allied Sciences* 69(2), April 2014, 221–50.

—— *The Trials of Psychedelic Therapy*, Johns Hopkins University Press, 2018.

Oreskes, Naomi, 'Objectivity or Heroism? On the Invisibility of Women in Science', *Osiris*, vol. 11: *Science in the Field*, University of Chicago Press, 1996, 87–113.

O'Shaughnessy, W.B., 'On the Preparations of Indian Hemp, or Gunjah', *Journal of the Asiatic Society of Bengal*, 1839, 838–51.

Owen, Alex, 'The Sorcerer and his Apprentice: Aleister Crowley and the Magical Exploration of Edwardian Subjectivity', *Journal of British Studies* 36(1), January 1997, 99–133.

—— *The Place of Enchantment: British Occultism and the Culture of the Modern*, University of Chicago Press, 2004.

Parascandola, John, 'The Drug Habit: The Association of the Word "Drug" with "Abuse" in American History', in *Drugs and Narcotics in History*, ed. Roy Porter and Mikuláš Teich, Cambridge University Press, 1997, 156–67.

Parssinen, Terry M., *Secret Passions, Secret Remedies: Narcotic Drugs in British Society 1820–1930*, Manchester University Press, 1983.

Partridge, Christopher, *High Culture: Drugs, Mysticism and the Pursuit of Transcendence in the Modern World*, Oxford University Press, 2018.

Pendergrast, Mark, *For God, Country and Coca-Cola*, Basic Books, 2013.

Perry, Ralph Barton, *The Thought and Character of William James*, Vanderbilt University Press, 1948.

Phillips, Adam, *Becoming Freud*, Yale University Press, 2014.

Pollan, Michael, *How to Change your Mind*, Allen Lane, 2018.

—— *This Is your Mind on Plants*, Allen Lane, 2021.

Preciado, Paul B., *Testo Junkie*, Feminist Press, 2013 [2008].

Putnam, Robert D., with Shaylyn Romney Garrett, *The Upswing*, Simon & Schuster, 2020.

Rabinbach, Anson, *The Human Motor: Energy, Fatigue and the Origins of Modernity*, University of California Press, 1990.

Ramsay, William R., 'Partial Anaesthesia', *Proceedings of the Society for Psychical Research*, 9, 1893, 236–44.

Randolph, Paschal Beverly, *Hashish: Its Uses, Abuses and Dangers, its Extasia and Fantasia, and Illuminati*, published with *Guide to Clairvoyance*, Rockwell & Rollins, 1867.

Rankin, Nick, *Dead Man's Chest: Travels after Robert Louis Stevenson*, Phoenix Press, 2001 [1987].

Rasmussen, Nicolas, *On Speed*, New York University Press, 2008.

Reed, Edward S., *From Soul to Mind: The Emergence of Psychology, from Erasmus Darwin to William James*, Yale University Press, 1997.

Regardie, Israel, *Roll Away the Stone, and The Herb Dangerous*, Newcastle Publishing, 1994 [1968].

Richards, William A., 'Abraham Maslow's Interest in Psychedelic Research: A Tribute', *Journal of Humanistic Psychology* 57(4), September 2016, 319–22.

Richardson, Benjamin Ward, 'On Ether-Drinking and Extra-Alcoholic Intoxication', *Gentleman's Magazine* 1774, 1893, 440–64.

Richardson, J., *Théophile Gautier, his Life and Times*, M. Reinhardt, 1958.

Richardson, Robert D., *William James in the Maelstrom of American Modernism*, Mariner Books, 2007.

Rieger, Christy, 'Chemical Romance: Genre and *Materia Medica* in Late-Victorian Drug Fiction', *Victorian Literature and Culture* 47(2), 2019, 409–37.

Rivers, W.H.R., *The Influence of Alcohol and Other Drugs on Fatigue*, Edward Arnold, 1908.

Roof, Wade Clark, *A Generation of Seekers: The Spiritual Journeys of the Baby Boom Generation*, Harper San Francisco, 1993.

Roszak, Theodore, *The Making of a Counterculture*, University of California Press, 1968.

Rushton, Sharon, *Creating Romanticism*, Palgrave Macmillan, 2013.

Russell, Bertrand, *A History of Western Philosophy*, Simon & Schuster, 1945.

St John, Graham, *Mystery School in Hyperspace: A Cultural History of DMT*, Evolver Editions, 2015.

Saler, Michael, *As If Modern Enchantment and the Literary Prehistory of Virtual Reality*, Oxford University Press, 2012.

—— (ed.), *The Fin-de-Siècle World*, Routledge, 2015.

Samorini, Giorgio, 'Paolo Mantegazza (1831–1910): Pionere italiano degli studi sulle droghe', *Eleusis* 2, September 1995, 14–20.

—— *L'erba di Carlo Erba*, Nautilus Press, 1996.

Sargant, William, *The Unquiet Mind*, William Heinemann, 1967.

Schaffer, Simon, 'Self Evidence', *Critical Inquiry* 18(2), Winter 1992, 327–62.

Schiebinger, Londa, *Plants and Empire: Colonial Bioprospecting in the Atlantic World*, Harvard University Press, 2005.

Schivelbusch, Wolfgang, *Tastes of Paradise*, trans. David Jacobson, Vintage Books, 1993.

—— *Disenchanted Night: The Industrialisation of Light in the Nineteenth Century*, University of California Press, 1995.

Schmied, Lori A., Hannah Steinberg and Elizabeth Sykes, 'Psychopharmacology's Debt to Experiential Psychology', *History of Psychology* 9(2), 2006, 144–57.

Schmitt, Natalie Crohn, 'Ecstasy and Peak-Experience: W.B. Yeats, Marghanita Laski, and Abraham Maslow', *Comparative Drama* 28(2), Summer 1994, 167–81.

Schmitz, Oscar A.H., *Hashish*, Wakefield Press, 2018 [1902].

Schultes, R.E., 'Plantae Mexicanae II: The Identification of Teonanacatl, a Narcotic Basidiomycete of the Aztecs', *Botanical Museum Leaflets, Harvard University* 7(3), 1939, 37–56.

Schuster, David G., *Neurasthenic Nation*, Rutgers University Press, 2011.

Sengoopta, Chandak, *The Most Secret Quintessence of Life*, University of Chicago Press, 2006.

Shapin, Steven, *The Scientific Revolution*, University of Chicago Press, 1996.

—— and Simon Schaffer, *Leviathan and the Air-Pump*, Princeton University Press, 1985.

Shattuck, Roger, *The Banquet Years*, Vintage Books, 1955.

Sheppard, Julia, *Silas Burroughs, the Man who Made Wellcome*, Lutterworth Press, 2022.

Sherwood, Alexander M., et al., 'Synthesis and Characterization of 5-MEO-DMT Succinate for Clinical Use', *ACS Omega* 5(49), 2020, 32067–75.

Shiel, M.P., *Prince Zaleski*, Martin Secker, 1928 [1895].

Shulgin, Alexander, and Ann Shulgin, *Pihkal: A Chemical Love Story*, Transform Press, 1991.

——, —— and Peyton Jacob, 'A Protocol for the Evaluation of New Psychoactive Drugs in Man', *Methods and Findings in Experimental and Clinical Pharmacology* 8(5), 1986, 313–20.

Siegel, Jerrold, *Bohemian Paris: Culture, Politics and the Boundaries of Bourgeois Life 1830–1930*, Johns Hopkins University Press, 1986.

Simner, Julia, and Edward M. Hubbard (eds), *The Oxford Handbook of Synesthesia*, Oxford University Press, 2013.

Small, Douglas, 'Masters of Healing: Cocaine and the Ideal of the Victorian Medical Man', *Journal of Victorian Culture* 21(1), 2016, 3–20.

Smith, W.D.A., *Under the Influence: A History of Nitrous Oxide and Oxygen Anaesthesia*, Wood Library-Museum of Anaesthesiology, 1982.

Snow, Stephanie, *Blessed Days of Anaesthesia*, Oxford University Press, 2008.

Solhdju, Katrin, *Travelling into Alienation: Moreau de Tours' Experimental Attempts to Articulate the Body of Madness*, published online at https://dingdingdong.org/wp-content/uploads/2012/12/alienatingtravels.pdf, 2012.

Sommer, Andreas, 'James and Psychical Research in Context', in *The Oxford Handbook of William James*, ed. Alexander Klein, Oxford University Press, advance publication online, 2020.

Spagnoli, Laura, *Under the Influence: Literature, Drugs and Modernity in France (1870–1914)*, dissertation, University of Pennsylvania, 2002.

Spillane, Joseph F., 'Making a Modern Drug: The Manufacture, Sale and Control of Cocaine in the United States 1880–1920', in *Cocaine, Global Histories*, ed. P. Gootenberg, Routledge, 1999, 21–45.

Stansfield, Dorothy, *Thomas Beddoes M.D. 1760–1808*, D. Reidel, 1984.

Stevens, Jay, *Storming Heaven: LSD and the American Dream*, Paladin Books, 1989.

Stevenson, Robert Louis, *Dr Jekyll and Mr Hyde*, Penguin Classics, 1979 [1886].

—— *Essays of Robert Louis Stevenson*, ed. William Lyon Phelps, Classic Literature Library, 1906.

Stewart, Omer C., *Peyote Religion: A History*, University of Oklahoma Press, 1987.

Stratmann, Linda, *Chloroform: The Quest for Oblivion*, Sutton Publishing, 2003.

Strickland, Stuart Walker, 'The Ideology of Knowledge and the Practice of Self-Experimentation', *Eighteenth-Century Studies* 31(4), Summer 1998, 453–71.

Symons, Arthur, *The Symbolist Movement in Literature*, Carcanet Press, 2014 [1899].

—— 'The Gateway to an Artificial Paradise: Hashish and Opium Compared', *Vanity Fair*, October 1918.

Synge, John M., *Under Ether*, in *The Works of John M. Synge*, vol. 4, John W. Luce & Co., 1912 [1897].

Tallis, Frank, *Hidden Minds: A History of the Unconscious*, Helios Press, 2012 [2002].

Taylor, Bayard, *The Lands of the Saracen*, G.P. Putnam, 1855.

Taylor, Charles, *Sources of the Self: The Making of Modern Identity*, Harvard University Press, 1989.

Teive, Hélio et al., 'Overcoming Bashfulness: How Cocaine Aided Freud to Summon the Courage to Meet Charcot', *Arquivos de Neuro-Psiquiatria* 7(11), November 2019, 825–7.

Timberlake, James H., *Prohibition and the Progressive Movement 1900–1920*, Harvard University Press, 1963.

Torgoff, Martin, *Can't Find my Way Home: America in the Great Stoned Age 1945–2000*, Simon & Schuster, 2004.

Travers, Morris W., *A Life of Sir William Ramsay*, Edward Arnold, 1956.

Turner, Fred, *The Democratic Surround: Multimedia and American Liberalism from World War II to the Psychedelic Sixties*, University of Chicago Press, 2014.

Tymoczko, Dmitri, 'The Nitrous Oxide Philosopher', *The Atlantic*, May 1996.

Uthaug, M.V. et al., 'A single inhalation of vapor from dried toad secretion containing 5-methoxy-N,N-dimethyltryptamine (5-MeO-DMT) in a naturalistic setting is related to sustained enhancement of satisfaction with life, mindfulness-related capacities, and a decrement of psychopathological symptoms', *Psychopharmacology* 236, 2019, 2653–66.

Villavicencio, Manuel, *Geografía de la República del Ecuador*, Imprenta de Robert Craighead 1858.

von Bibra, Baron Ernst, *Plant Intoxicants*, Healing Arts Press, 1995 [1855].

Warren, John Collis, *Effects of Chloroform and of Strong Chloric Ether, as Narcotic Agents*, William D. Ticknor & Co., 1849.

Weinberg, Bennett Alan, and Bonnie K. Beaker, *The World of Caffeine*, Routledge, 2001.

Wells, H.G., *The Complete Short Stories of H.G. Wells*, A. & C. Black, 1927.

Whorton, James C., *The Arsenic Century*, Oxford University Press, 2010.

Wilson, Frances, *Guilty Thing: A Life of Thomas De Quincey*, Bloomsbury, 2016.

Winchester, Jake, 'Solanaeous Fumigations in Nineteenth-Century London', *Psychedelic Press Journal* 28, 2019.

Witkiewicz, Stanisław Ignacy, *Narcotics*, trans. Soren A. Gauger, Twisted Spoon Press, 2018 [1930].

Wyld, George, *Theosophy and the Higher Life*, Trübner & Co., 1880.

—— *Notes of my Life*, Kegan Publishers, 1903.

Yeats, W.B., *Discoveries: A Volume of Essays*, Dun Emer Press, 1907.

—— *The Trembling of the Veil*, T. Werner Laurie Ltd, 1922.

Young, James Harvey, *Pure Food: Securing the Federal Foods and Drug Act 1906*, Princeton University Press, 1989.

Zaehner, R.C., *Mysticism: Sacred and Profane*, Clarendon Press, 1957.

ACKNOWLEDGMENTS

I'd like to pay tribute to two valued companions on this journey, Michael Neve and Nick Tosches, who died within days of each other in October 2019. It was Michael who first encouraged me to study the history of drugs and introduced me to the library at the Wellcome Trust Centre for the History of Medicine, where he taught. Many of the characters and episodes in this book took shape during twenty years of conversations with him. I corresponded with Nick for several years, mostly about James Lee, whose life and times he researched with characteristic tenacity. Working alongside him was an education.

My thanks to Toni Melechi for many years of stimulating conversation and correspondence about the history of psychology and self-experiment, and for his comments on the draft manuscript. Thanks also to my other early readers: John Tresch, for his wide-ranging insights and suggestions; Andrew Lees, for casting an expert eye over the neurology and pharmacology; and Peter Moore, who kept the conversation flowing over outdoor pints between lockdowns.

Thanks too to the many people who assisted me with research queries and archival access during pandemic shutdowns, including David Barrett; Ben Breen; Alastair Brotchie; Rob Dickins; Adam Green at Public Domain Review; David Guba; Phoebe Harkins and Ross Macfarlane at the Wellcome Library; Peter Johnson, Secretary of the Society of Psychical Research; Axel Klein; Sarbajit Mitra; Jill Moretto, archive manager at GSK; Sujaan Mukherjee; Mike Power; James Pugh; Keeper Trout and A J Wright. I'm very grateful to Sonu Shamdasani at UCL's Health Humanities Centre for facilitating an Honorary Research Fellowship that allowed me remote access to academic journals while libraries were shuttered. Special thanks, as always, to Lou for making the process as easy and enjoyable as possible.

Thanks once again to my agent, Caroline Montgomery, and to the editorial and production staff at Yale for realising this project so smoothly and to such a high standard. I'm particularly grateful to my editor Julian Loose, whose expertise was invaluable in shaping the proposal and guiding the book to its final form.

INDEX

5-MEO-DMT, 2, 13; *see also* bufotenine

absinthe, 152, 255, 263
Academy of Medicine, France, 193
Academy of Sciences, France, 117
addiction, 4, 5, 10, 49, 68–72, 73, 79, 81,
   84, 86, 122–3, 246, 256–8, 274, 312,
   314; among doctors 70, 84, 122;
   'cocainomania', 72, 79; in Thomas De
   Quincey, 49; 'narcomania', 69, 246
Adler, Alfred, 296
adrenaline, 98, 285, 329n.
alcohol, 10, 25, 26, 29, 33, 36, 58, 68, 81,
   198, 242–3, 245, 249, 250, 254, 256,
   268, 276, 294; alcoholism, 29, 28, 68,
   81, 250, 256; medical uses, 25;
   prohibition, 242, 273
Alexander, Michelle, 331n.
alkaloids, 30, 45, 67, 75, 168, 230
Allbutt, Clifford, 256
Alles, Gordon, 284–5, 305
Alpert, Richard (Ram Dass), 282, 286, 287,
   301–2
American Jewish Committee, 279
American Medical Association, 248, 265
American Pharmaceutical Association, 242,
   331n.
American Psychological Association, 259
amphetamine, 98, 268, 282, 284–6, 292,
   300, 313; 'amphetamine psychosis', 300,
   mechanism of action, 313
amyl nitrite, 249
anaesthetics 8, 63–7, 71, 106, 115–21,
   122–6, 133, 134, 138, 140–3, 147–8,
   170; 'anaesthesia' coined, 124; anomalous
   experiences under, 106–7, 122–3, 126,
   140–3, 147–8; cocaine as local
   anaesthetic, 63–7; dental anaesthesia,

106, 116; *see also* chloroform, cocaine,
   ether, nitrous oxide
Ancient Mesoamerica (journal), 1, 3
Andersen, Hans Christian, 121
Anslinger, Harry, 273–4, 276, 280
Apollonius of Tyana, 214–15
Arabian Nights, The, 167, 171–2, 179, 190,
   192, 207, 208, 209, 234, 236, 276; 'Tale
   of the Hashish Eater', 171
Arnold, Matthew, 10
Artaud, Antonin, 330n.
arsenic, 6, 26, 27, 329n.
Aschenbrandt, Theodor, 28–9, 268
Assassins (Hashishin, Nizari Isma'ilis),
   172–4, 186, 188, 190, 193, 209, 236,
   273–4, 282
asthma, 79, 152, 215, 225, 329n.
Atlantic Monthly, The, 163
atropine, 169
Aubert-Roche, Louis, 176–7, 180, 184,
   193
Authoritarian Personality, The (1950),
   279
'automatisms', 145
ayahuasca, 280, 301, 330n.

Bakhtin, Mikhail, 326n.
Balzac, Honoré de, 25–6, 50, 189; and the
   Club des Hashischins, 189; self-
   experiments with coffee, 25–6; *Treatise on
   Modern Stimulants* (1839), 25
Banner of Light, 211
barbiturates, 266, 294
Bartholow, Philips, 72
Barton, William, 110
Baudelaire, Charles, 6, 51, 87, 152, 189,
   192, 194–9, 208–9, 218, 224, 225, 227,
   255, 261; and synaesthesia, 197, 228; *Les*

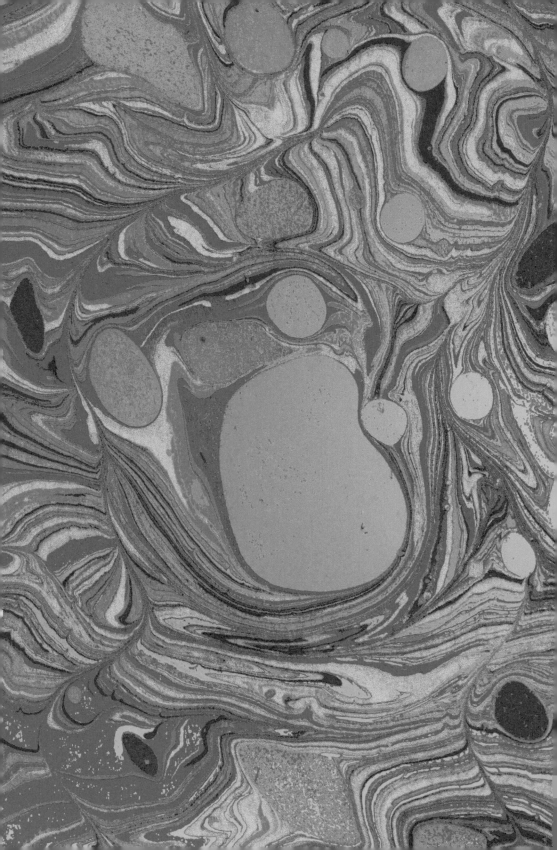